ha NG ON'

TO THE STAR.

STEPPING STONES TO THE STARS

THE STORY OF MANNED SPACEFLIGHT

TERRY C. TREADWELL

For my Grandson, Rex Treadwell

Front: (top, spine, back left) Views of the extravehicular activity during STS 41-B, courtesy of NASA; (bottom) Apollo -- July 1971, courtesy of NASA. *Back*: Bruce McCandless, courtesy of NASA.

First published 2010

The History Press
The Mill, Brimscombe Port
Stroud, Gloucestershire, GL5 2QG
www.thehistorypress.co.uk

British Library Cataloguing in Publication Data.
A catalogue record for this book is available from the British Library.

ISBN 978 0 7524 5409 2

Typesetting and origination by The History Press
Printed in Great Britain
Manufacturing managed by Jellyfish Print Solutions Ltd

CONTENTS

ACKNOWLEDGEMENTS

I would like to thank my wife Wendy for her time and patience in editing the manuscript and for her help and support. Thanks also go to Colonel Henry (Hank) Hartsfield, USAF (Ret.); Colonel Vladimir Kondratenko, former commander of the Russian Test Pilot School at Zukovsky; and Captain Bruce McCandless, USN – NASA Astronaut.

FOREWORD

One of the most fascinating subjects for me has always been the history of rockets and space exploration, especially as it relates to human spaceflight. As a young person I read science fiction and dreamed of being able one day to fly in space. Of course at that time I had little appreciation of the risks of this type of adventure and how things that appeared to be simple could be so dreadfully complicated. I never outgrew this deep interest in spaceflight, and then on 27 June 1982 my dream came true when I stepped aboard the Orbiter Enterprise (STS-4). This was followed by flights aboard STS-41D and STS-61A, both as commander. I retired from NASA in 1998 but today continue to work in the space industry.

There have been many books written about spaceflight covering many different facets of the history, such as the people involved, the development of the technology and theory, the development of the spacecraft, the details of specific flights, the glory, the failures – the list is long. Many books contain great detail about their specific subjects and sometimes are difficult to read. Since I am technically oriented, books containing great detail about the engineering evolution and solutions of great technological challenges and problems tend to attract my attention. Stories about the people involved are also interesting to me. But not everyone has this depth of interest. It seems to be that there is a lack of a good chronology of the development of space technology written for the lay person.

In this book Terry Treadwell has filled that void. He has accurately presented the essence of the development of space travel and concisely put it into print. His manner of writing has captured the excitement of the time with just enough detail to help the reader understand the significance of the events described without getting bogged down in great detail. The story told here is factual and sprinkled with some human details not often seen in books on this subject. The book begins with a brief history of experiments in reaction force propulsion by the Greeks and rockets by the Chinese, followed by rockets through the ages. Ultimately the work of Goddard and Tsiolkovsky pave the way toward the development of exoatmospheric rockets; then follows the gripping saga of human spaceflight in the twentieth century, leading into the new millennium. This history is complete in that it discusses chronologically step by step the progress, the joy, and sometimes tears, experienced by the major space-fairing nations, the United States and Soviet Union/Russia, in the conquest of space.

Terry has often written about aviation and possesses a keen interest in its history. For the twenty plus years I have known him, he has also been fascinated by space travel and has now turned his interest and writing skills to this topic with excellent results.

Colonel Henry W. Hartsfield Jr. NASA Astronaut

Member of the 2nd Group USAF Manned Orbiting Laboratory – 1966.
Selected for NASA-Group 7 – 1969.
Member of Support crew for Apollo XVI.
Member of Support crew for Skylab 2, 3 & 4.
Back-up pilot STS-2.
Back-up pilot STS-3.
Pilot of STS-4 – Columbia.
Commander of STS-41D – Discovery (Maiden Flight).
Commander of STS-61A (Spacelab D-1) – Challenger.
Resigned from NASA in 1998 to join RAYTHEON in developing the simulators for the International Space Station (ISS).

INTRODUCTION

This book is essentially a pictorial history of the development of manned spaceflight. Included is the first manned rocket flight that took place in 1933, the launch of the first satellite, Sputnik, followed closely by the first manned flight into space by Russian cosmonaut Yuri Gagarin and the beginning of the space race to land a man on another world.

There are accounts of the tragedies that struck: the deaths of the three American astronauts and the lessons learned from them, and the deaths of the Russian cosmonauts in their desire to get into space and closer to other worlds.

All this was closely followed by the development of the Russian Space Stations and the strides made in preparing man to live and work in space. This culminated in the development of the first re-usable spacecraft – the Space Shuttle. Also covered are the early design years and the variety of proposals from different companies like Grumman and Boeing, before Rockwell was selected as the prime contractor for the Space Shuttle.

The successes of this complicated rocket/glider are highlighted, as are the failures. The potential that this system has to offer is apparent when the various payloads that were carried into space are mentioned. Covered briefly is the Soviet STS System with its 'Buran' Orbiter and its failure to turn the project into a viable alternative to the rocket.

Some of the photographs in the book have never been seen before and are some of the most dramatic and breathtaking. A number of cutaway drawings of both the Russian and American spacecraft support the information provided in the text and photographs. The book is not meant to be the definitive work on this vast and complicated subject, but it will give a photographic insight into one of the last millennium's greatest achievements, which enabled men and women to work in an environment outside of the Earth.

CHAPTER ONE

THE BIRTH OF THE ROCKET

When man first gazed at the stars and dreamed of going to other worlds it all looked more than a million miles away, as indeed it was. The foundations of astronomy were laid by Greek scientists around 624 BC, when the teachings of Thales of Miletus were recorded. They realised that the Earth was just a tiny speck in the universe and man was just an infinitesimal part of the whole thing. Then as the years progressed and man took to the air, first in balloons, then gliders and later engine-powered flying machines, the dreams started to take on a different light, becoming reality.

The first recorded use of a device that was used in propulsion was by way of a demonstration in AD 160, by a Greek mathematician and scientist called Hero of Alexandria. The device, called an Aeolipile (Greek for 'ball' and after Aeolus, the god of winds), consisted of a rotating hollow sphere with two right-angle pipes located 180° apart, which were mounted between two supports that carried steam from a closed container suspended over a fire. The jet of steam escaping through the pipes caused the sphere to revolve and is the first known demonstration of a crude jet propulsion system.

Then in AD 1232, during the Mongol siege of the city Kai-fung-fu, the first recorded use of a rocket was identified in the Chinese chronicle *T-hung-lian-kang-mu*. The 'Arrow of Flying Fire', as it was called, was most likely to have been a hollow arrow that was stuffed full of some kind of incendiary material, or an arrow that had been attached to a similar substance. According to one report the rockets were very large and, when lit, made a noise that resembled thunder and could be heard over a distance of 5 leagues (15 miles). When falling to Earth, it was said to have caused devastation for 2,000ft in all directions. Reports also stated that the rockets carried iron shrapnel and incendiary material.

The first recorded use of rockets in warfare in Europe was in AD 1241, when the Mongols used them with devastating effect against the Magyar forces in the Battle of Sejo, which resulted in the capture of Buda – now known as Budapest.

In Europe around 1249, an English monk by the name of Roger Bacon (1214–94) is credited with developing the first rockets in the Western world. Bacon was a philosopher and scientist and foresaw the use of gunpowder, together with mechanical cars, boats and planes. In his book *De Mirabili Potestate Artis et Naturae* (The Miracles of Art, Nature and Magic) he gave the formula for gunpowder as follows:

> Take 7 parts of saltpeter, 5 of young hazel twigs and 5 of sulphur, and thou wilt call up thunder and destruction, if thou know the art.

The Mongols used the rocket again in 1258 when they launched them against Arabs during their capture of Baghdad (750 years later rockets were once again used by the British and Americans in the battle for Baghdad). Ten years later, in 1268, the Arabs, quick to learn, used the rocket against Louis IX during the Seventh Crusade.

By the year 1300 a number of European armies had acquired rockets and formed rocket corps. In 1429 the French used rockets at the Siege of Orleans during the Hundred Years War.

In Russia, however, virtually no records exist of the use of rockets until the 1600s, when accounts accumulated by the Russian gunsmith Onisin Mikhailov were turned into a compiled document entitled *Code of Military, Artillery and Other Matters Pertaining to the Science of Warfare*. In this document detailed descriptions of rockets referred to as 'Cannon balls which run and burn' are contained, although now even the existence of this early document has been disputed, mainly because the information which compiled the main manuscript was not entirely Mikhailov's. It was, in fact, a collection of some 663 snatches of information and articles from a variety of foreign military books and sources. But that doesn't really matter; the fact that Mikhailov collected and published these articles is of more importance than the debate to try and decide whether or not he was justified in calling them his own.

In 1650 the Dutch were using military rockets during their wars, followed closely by the Germans, who in 1730, under the control of Colonel Friedrich von Geissler, started manufacturing rockets that weighed 120lb.

Peter the Great of Russia devoted a lifetime to creating his country's military might. In 1680 he founded the first Rocket Works in Moscow. There they made illuminating and signal rockets for the army, all of which were under the guidance of English, Scottish, Dutch, German and French officers, who instructed them in their use. Then in the early 1700s Peter the Great made St Petersburg the new capital of Russia and moved the entire Rocket Works to this new location, expanding it at the same time.

The use of the rocket continued to be developed by various nations, but it really made its mark in warfare during the Indian campaign at the end of the eighteenth century. In 1789 an Indian force of 1,200 men under the ruler Hyder Ali, Prince of Mysore, devastated the British army at the Battle of Panipat using iron rockets. They inflicted horrendous casualties, despite the fact that the British far outnumbered and outgunned them. The rocket was 8in long, 1.5in in diameter and was fixed to a stabilising 8ft rod.

Three years later, in 1792, the son of Hyder Ali, Tippu Sultan, enlarged the rocket corps created by his father to 5,000 men and supplied them with larger rockets. During the Third Mysore War he inflicted several defeats upon the British, before being killed at the Battle of Seringapatam.

Because of the devastation caused by these rockets, Sir William Congreve carried out secret experiments at the Royal Laboratory at Woolwich, England, in developing substantially more powerful ones. After two successive campaigns, (Boulogne in 1806 and Copenhagen in 1813) where the Danes were subjected to a barrage of some 25,000 rockets, the British army formed the Field Rocket Brigade. Also about that time, Lieutenant-General Henry Shrapnel invented the explosive shell, which caused serious casualties to Napoleon's cavalry and close-quarter infantry. During the battles against Napoleon, the rocket brigade was involved in every campaign, distinguishing itself particularly well in the final battle at Waterloo when Napoleon was trounced.

The development of the Russian rockets continued, and in the late 1700s an officer in the tsar's artillery, Alexander D. Zasyadko, who had been studying Congreve's progress and exploits, together with the files from the Rocket Works, decided to design some rockets of his own. So successful were the tests of his rockets that in 1817 Zasyadko was assigned to western Russia to train the tsar's soldiers in the use of military rockets. The following year a school of artillery was opened and Zasyadko was appointed its head with a promotion to major general. In the Russo-Turkish War of 1828–29, solid fuel rockets were used during the sieges of Varna, Braila, Silistra and Schmia, and on the Black Sea Russian ships used the rockets with great success. During the Crimean War thousands of rockets were employed, with increasing reliability.

During the American War of Independence, the British army created a rocket brigade after the Battle of Bladensburg on 24 August 1814 when the British 85th Light Infantry used rockets against an American rifle battalion. According to a report by Lieutenant George Gleig of the 85th:

> Never did men with arms in their hands make better use of their legs.

The development of the rocket in warfare moved on in the 1840s, when an Englishman, William Hale, developed a rocket with three curved metal vanes in the exhaust creating the first stabilised rocket. The United States employed it to great effect in the Mexican War of 1846–48. A battalion of rocketeers, consisting of about 150 men armed with around fifty rockets, and under the command of

1st Lieutenant George H. Talcott, accompanied Major-General Winfield Scott's army. During the battle for Veracruz on 24 March 1847, they were used for the first time and with great success. With Veracruz taken, the rocketeers were moved to the battle for Telegraph Hill, where Captain Robert E. Lee was in command. Lee was later to command the Confederate Army of Northern Virginia in the American Civil War. Mexico City surrendered some months later when the fortress of Chapultepec fell after heavy bombardment from rockets and a massive assault.

On the other side of the world in Russia, the death of Zasyadko in 1837 caused his position at the school to be taken by another artillery officer by the name of Konstantin I. Konstantinov. The 30-year-old officer was the first to work on the practical problems of rocket production and became the founder of experimental rocket dynamics. Up to this time, the production of rockets was left to the individual skill of the makers and, as can be imagined, there were a number of accidents. Among Konstantinov's achievements were the development of large-scale rocket production and a rocket that could fire lifelines to wrecked vessels.

A number of other Russian inventors produced ideas over the next few years, and then came the innovations of Konstantin Eduardovich Tsiolkovsky, considered by the Russians to be the 'Father of Soviet Spaceflight'. Born in 1857, the son of a forestry expert and inventor, Tsiolkovsky enjoyed a normal childhood. Then, when he was 8 years old, he contracted scarlet fever and became almost totally deaf. Unable to go to school, he taught himself from his father's books, mastering first mathematics then physics. After three years at a technical school he returned to his hometown to become a teacher. It was whilst he was a teacher that he started to carry out serious research in the areas of an airplane, an all-metal dirigible and a rocket for interplanetary travel.

This was endorsed in 1881, when Nikolai I. Kibalchich proposed the idea of heavier-than-air machines being propelled by rocket propulsion and carrying human passengers. Unfortunately Kibalchich also used his expertise in another direction. He was the bomb expert for a revolutionary organisation, and a bomb thrown at the Tsar Alexander II, which mortally wounded him, had been prepared by Kibalchich. He was arrested by the tsar's secret police and executed soon afterwards.

Tsiolkovsky was a man of vision and was years ahead of his time in his concepts and designs. Between 1885 and 1895, Tsiolkovsky carried out a great deal of work on the design of metal airships although at first he was considered to be an eccentric living in a world of fantasy. In 1894 he proposed a design for an all-metal airplane in an article entitled, 'The Airplane, a Birdlike Flying Machine'. Seven years later he put forward a paper 'Investigating Space with Reaction Devices'. His work was slowly becoming accepted and he was no longer regarded as an eccentric, but it wasn't until 1918 that he was properly recognised for what he had achieved. In 1919 he was elected to the Russian Socialist Academy and granted a personal pension by the Commission for Improvement of the Lot of Scientists

(TsEKUBU). This later became the Academy of Sciences USSR. Tsiolkovsky died in 1935 a national hero, bequeathing all his papers and models to the Soviet government. In 1952, the Aero Club of France had a large gold medal struck in his honour and in 1954 the Soviet government established the Tsiolkovsky Gold Medal, which has been awarded every three years since its installation to the most outstanding contributor to spaceflight.

On the other side of the Atlantic, a Massachusetts physics teacher by the name of Robert Hutchings Goddard had never heard of Tsiolkovsky when he started taking an interest in rocketry at the beginning of the 1900s. In 1920, Goddard submitted a sixty-nine-page paper called 'A Method of Reaching Extreme Altitudes' to the Smithsonian Institute in Washington. The Smithsonian, recognising that there was a great deal of merit in the paper, gave Goddard a $5,000 grant towards further research. Unfortunately the story was leaked to the newspapers and some of the more sensation seeking tabloids of the time, always on the lookout to ridicule anything that they didn't understand, had a field day. This caused Goddard to continue his research in complete seclusion.

On 16 March 1926, Goddard launched the first liquid-fuelled rocket in history, although it only travelled to a height of 184ft. Even then the newspapers mocked him by bannering the headline 'Moon Rocket Misses Target by 238, 799.5 miles'. Later Goddard was supported by Charles Lindbergh and financed by the Guggenheim Foundation. By the end of 1935, Goddard's rockets had gyroscopic control, and were followed two years later by a rocket that reached a height of 1.9 miles.

Then, in 1933, an article appeared in a London newspaper, said to have been written by the special correspondent of a newspaper called the *Sunday Referee*, about a 'Sensational secret demonstration that was carried out on the island of Rugen in the Baltic Sea'. It was said that a German scientist by the name of Dr Bruno Fischer placed his brother Otto into a 24ft steel rocket and shot him 6 miles into the air. The whole experiment was conducted under the control of the German War Ministry (*Reichswehr*) and under great secrecy. Otto Fischer, it was said, crawled through a small steel door in the side of the rocket and strapped himself in. Minutes later there was a blinding flash followed almost immediately by a deafening roar and the steel rocket was blasted into the sky. Some minutes after that it came back into sight, floating down on a large parachute and came to rest on the sands. The occupant (one hesitates to call him a pilot) crawled out to the relief of the assembled government witnesses. The flight had lasted just 10 minutes and 26 seconds and had reached a height of 6 miles – the first recorded manned rocket flight is said to have taken place. There has always been some dispute over whether the story was a hoax or a propaganda exercise by the German authorities, or in fact actually did happen.

Germany's interest in rockets and space travel had started seriously in 1923, when a German-speaking Transylvanian schoolteacher by the name of Hermann

Group of German rocket scientists with a young Werner von Braun.

Oberth had his book, *Die Rakete zu den Planetenräumen* ('The Rocket into Interplanetary Space'), published. The book attracted a number of similar minded enthusiasts, and they arranged to meet in the backroom of a restaurant in Breslau called the Golden Sceptre. From this meeting came the society *Verein für Raumschiffahrt* (VfR – Society for Space Travel), whose aims were to promote the theory of space travel and carry out serious experiments in rocket propulsion. Within two years the society's membership had grown from a mere handful of enthusiasts to 870 dedicated members. Among these was a young 19-year old by the name of Werner von Braun.

Across the Atlantic, Goddard had continued his experiments in New Mexico, well away from the prying eyes of the press. His work was slow and ponderous at this stage, mainly because of the lack of interest in rocketry within the United States and, indeed, in the United Kingdom. This was highlighted in an article in the *Journal of the British Interplanetary Society* in 1937, which said:

> Astronautical activities seem to have ceased around the world lately, nothing having been done in the USA, Germany or Austria.

The outbreak of the Second World War seemed to put the experiments with rockets on hold, but four years after the article had appeared in the *British Interplanetary Journal*, a speech by Adolf Hitler in Danzig on 19 September 1941 included the following statement:

> If the war continues four or five years, Germany will have access to a weapon now under development that will not be available to other nations.

The was a worrying factor for the Allies, because up to then it had been thought that Germany, one of the leaders in the development of rockets, had shelved any interest in the subject. They were obviously mistaken as Britain found out to her cost. Hitler had kept all development secret, even though he had publicly denied any interest in the use of rockets. It is accepted that at the time there was a distinct disinterest among the military hierarchy, who regarded it as a complete waste of money and resources. Among these was Field Marshal Erwin Rommel.

Just prior to the Second World War, the *Luftwaffe* had made progress in leaps and bounds in the field of rocket science, developing first the A-1 and then the A-2, followed by the A-3 and A-4. It has always been thought that the development of these early rockets was primarily for use on a war footing, but this is not true. Financial backing for these experiments disappeared after the failure of the A-4, and Captain (later General) Walter Dornberger, who was in charge at the time, was desperate. Then, during the Second World War, funding suddenly became available when it was realised that the rockets had a military purpose. The research acquired from these early rocket trials resulted in the development of the V-1 (*Vergeltungswaffe* 2 -Retaliation Weapon 2), then the V-2, formerly the A-4. The V-1 was much like a rocket-powered aircraft, and in fact a number of test flights were carried out with pilots aboard. Among these pilots was Hannah Reitsch, Germany's premier female test pilot and one of the very few to survive a test flight in the V-1. The V-2, on the other hand, looked like a rocket should and was the forerunner of the American space programme.

The most significant breakthrough for the Germans came on 3 October 1942, when Major-General Walter Dornberger, commanding officer of the Peenemünde Army Experimental Station, gave the order to launch another V-2 rocket. The two previous attempts had resulted in disaster as neither had lifted off the pad before exploding. The countdown started as film cameras whirled. The

V-2 lifting off during a test flight from Peenemunde.

tall slender body of the rocket, lacquered black and white, gleamed in the sunlight as it slowly lifted off the pad followed by an immense roar as the 650,000hp rocket motor burst into life. As the rocket sped faster and faster into the clear blue sky, on the ground, scientists and engineers watched with growing confidence. The countdown came relentlessly over the loudspeakers '291-292-293'. The rocket was now over 6 miles high and travelling at over 3,000mph. It was now estimated that it would soon re-enter the Earth's atmosphere and the speed of the rocket would decrease to 2,000mph. Then came the announcement over the loudspeakers: 'Impact'. The rocket had come to Earth and had landed with an impact of 1.400 million ft-lb.

Dornberger was delighted. The team, headed by Werner von Braun, had proved that they had the capability of launching ICBMs (Inter-Continental Ballistic Missiles) and had justified the funding that he had fought tooth and nail to obtain. Fortunately for Britain, it was to be a further two years before Hitler was to sanction their use; the damage the V-2, as it became known, caused to Britain is well documented. The German scientists, however, still had ideas of launching their rocket into space and, with this in mind, they continued to carry out experiments. Werner von Braun made no secret of the fact that he despised the Nazi regime, especially the secret police. This was a dangerous game to play; Heinrich Himmler had von Braun arrested on a number of occasions on charges of not doing enough to aid the German war effort. It was only the repeated intervention of General Dornberger that saved the scientist's life.

America's entry into the world of space exploration was given an unexpected boost toward the end of the Second World War, and started rather ignominiously on a road to a small village called Schattwald, close to the German/Austrian border. An American soldier with the German name of Schneikert was on guard duty with members of the 324th Infantry Regiment, when a smartly dressed man approached them on a bicycle. Schneikert, remembering some German from his childhood, called out '*Komm vorwärts mit die Hände Hoch!*' (Come forwards with your hands up). The young German stopped, got off his bicycle and stepped forward with his hands in the air. Speaking in a mixture of German and English, he explained that his name was Magnus von Braun and that his brother was Werner von Braun, a rocket scientist, who together with a number of other scientists, wanted to surrender to the Americans. After a great deal of suspicious mutterings between them, the American soldiers took Magnus von Braun to a member of the CIC (Counter Intelligence Corps), 1st Lt Charles L. Stewart, who established that he was indeed who he said he was.

It appeared that Werner von Braun, and over 100 of Germany's top rocket scientists and engineers, were hiding in a mountain inn, and were being hunted down by a SS group under the command of Obergruppenführer General und Waffen SS Hans Kammler, and it was imperative that the Americans got to them first. Kammler had been in charge of security at Peenemünde, and when it was obvious that the war was lost, had been given orders to kill all rocket scientists and engineers, to prevent them falling into the hands of the Allies. Stewart decided that the man was telling the truth, but was not prepared to risk any of his men to go and fetch the scientists. Instead he gave Magnus von Braun passes for the scientists to allow them through the American-held sector and told him to bring the party to him, guaranteeing their safety when they reached the American lines. The following day the party arrived at Stewart's headquarters and was taken into 'protective custody'.

At the end of the war the Americans raced to capture the remaining German scientists and rocket engineers in order to use their expertise for military and space exploration purposes. The majority of the German rocket scientists, including the top scientist Werner von Braun, were whisked away to America to set up the American rocket system under the control of the NACA (National Advisory Committee for Aeronautics). In America this became known as the 'Paper Clip' operation. This initially caused some concern amongst the Allies as a number of the scientists were known to be fanatical Nazis, causing some cynics to call it the 'Paper Clip Conspiracy'. The remainder of the German scientists were rounded up and taken to Russia to help in the Soviet rocket programme.

The majority of Goddard's work didn't come to light until the late 1940s, by which time the German rocket scientists had been 'liberated' from Europe and Robert Goddard had died. This, in real terms, meant that the Americans had no one of their own working seriously on rocket development.

Although they were allowed to enter the United States, the German scientists and engineers were kept under strict control. The group was sent to White Sands Proving Grounds, New Mexico, but it was to be a further two years before their families were able join them. It was suddenly realised that although the families had been granted entry visas, the original group of scientists and engineers had been taken to the United States by the military and had no entry visas. They were, in effect, illegal immigrants. This problem was solved by taking the unusual step of bussing the group from El Paso, Texas, to Ciudad Juárez, Mexico, then allowing them entry from Mexico into the United States and giving them visas at the same time.

With the German immigrants settled into their new home at White Sands, the US government decided that the honeymoon period was over and it was time to set them to work. Work started in assembling and testing one of the V-2 rockets that had been brought from Germany. This was not as straightforward as was first thought. The rocket components were collected from various parts of Germany and shipped to White Sands Proving Ground, but only two or three were assembled from original matching parts. Some parts had deteriorated so badly during transit that new parts had to be machined to replace them.

The first launch was a disaster. The rocket reached a height of 3.5 miles before one of the stabilising fins came off and the rocket had to be destroyed. Fortunately the second launch, attended by a large number of senior military officers, was a complete success and reached a height of 71 miles. There were a number of accidents during this learning period, one of which nearly caused a disaster. This happened when the gyroscope in a guidance unit malfunctioned and the rocket went haywire. It went seriously off course and headed for the town of Juárez in Mexico. Fortunately it missed the buildings where construction companies in the area stored their dynamite, and crashed close to the town. Within hours, the locals had set up stalls and were selling still warm remnants of the rocket as souvenirs. It was estimated later that the Mexicans sold about 10 tons of material, much of which resembled tin cans, all from a rocket only weighing 4 tons.

It was in 1949, during the tests of the V-2, that the first experiments using animals were conducted. Two monkeys named Albert I and Albert II were placed in the nose cones of the V-2s during launch tests. These tests were followed by a series of high altitude flights carrying Albert I and II again and III and IV. All of these animals died when the parachutes on the rockets failed to open during re-entry. A number of other flights carried a variety of creatures including mice, insects and cats.

Between April 1946 and September 1952 a total of sixty-four V-2 rockets were launched from the White Sands Proving Ground, New Mexico, and one was even fired from the deck of the aircraft carrier USS *Midway*.

Over the next few years the development of the rocket slowly gathered pace, but even then the scientists and engineers were not ready for the sudden outburst of frenzied activity in space and missile development that was to arrive in the late 1950s.

CHAPTER TWO

THE FIRST MANNED SPACEFLIGHTS

On 4 October 1957, the steppes of Kazakhstan in the Soviet Union trembled as a Vostok rocket was blasted into space. Inside was a metal ball with a diameter of just 23in and weighing 184lb, fitted with a radio transmitter. Minutes later a faint bleeping sound was tracked around the world as Korabl Sputnik 1, as the tiny object was called, circled the Earth. Man had entered space and the race to attempt to explore this element had started in earnest.

One month later, on 3 November 1957, another Vostok rocket carrying Korabl Sputnik 2 blasted off and the first passenger to orbit the Earth was launched into space. The passenger was a dog named Laika. Laika was a 3-year-old mongrel bitch that had been found scavenging in the streets of Moscow. She was one of three dogs to be trained for space related tests; the other two, Albina and Mushka, carried out flights on high-altitude rockets and instrumentation and life support tests respectively.

Laika was placed inside the nose of the satellite three days before the launch, with warm air being pumped in to keep her temperature constant. Just prior to the launch she was groomed and sensors placed on her body so that her bodily functions could be monitored. During the first few minutes of launch her heartbeat doubled; this settled down after about 3 hours but problems with the separation part of the core section that contained the thermal control system caused the temperature inside the cabin to increase to 40°C (104°F). The stress the dog was suffering was now becoming obvious and although it was first said that she spent seven days in orbit before being put to sleep by means of an automatic lethal injection, it is now known she died just hours after the launch from heat and distress after part of the nose cone was ripped off during the launch.

Prior to the launch of Laika, some thirty-four dogs on seventeen sub-orbital missions had been carried out and all but six dogs were recovered successfully.

The Americans, initially stunned at the sudden progress made by the Russians, stepped up their development programme and launched their first artificial satellite Explorer 1 into orbit on 1 February 1958. It weighed 8.3kg and had a

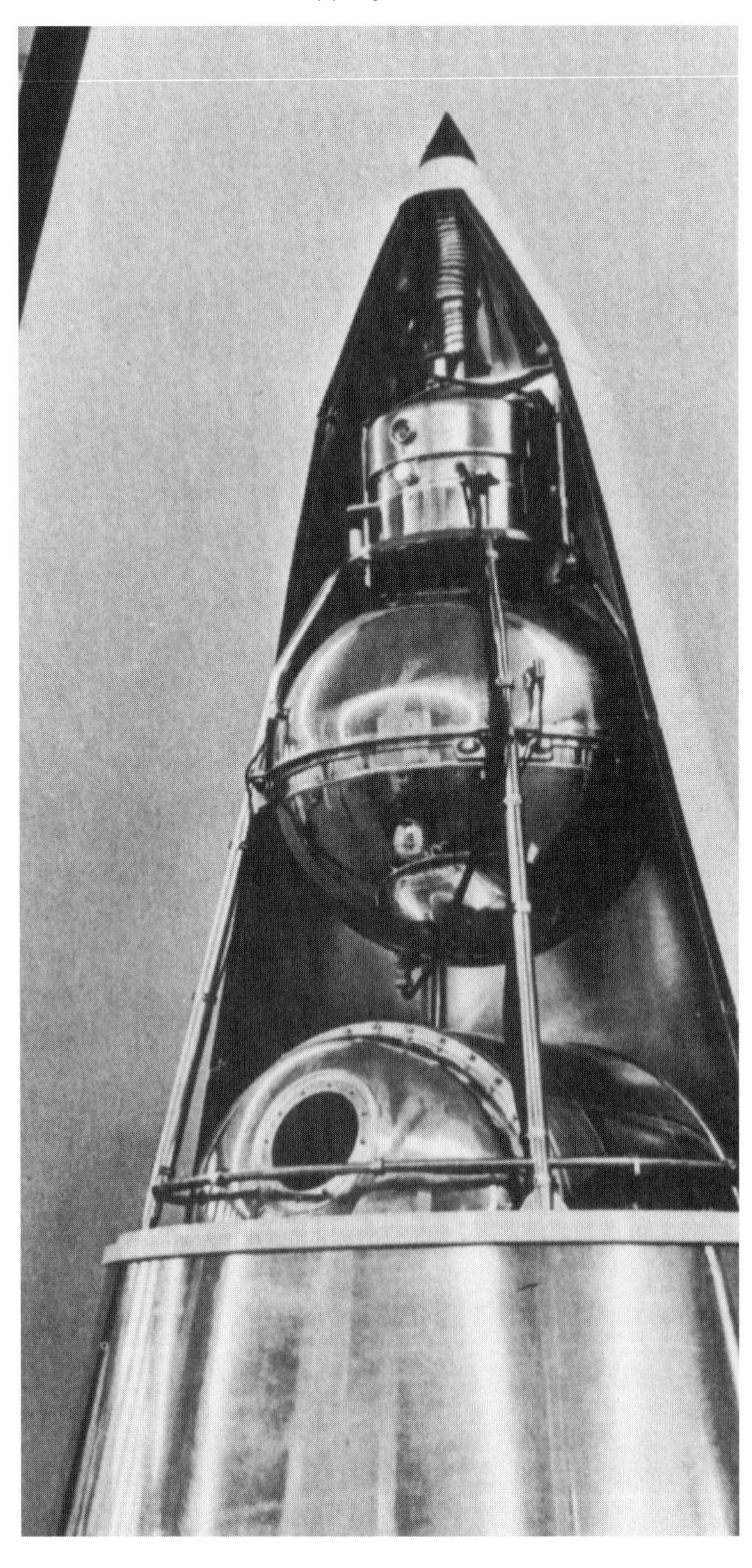

Capsule in which the dog Laika was placed for her spaceflight.

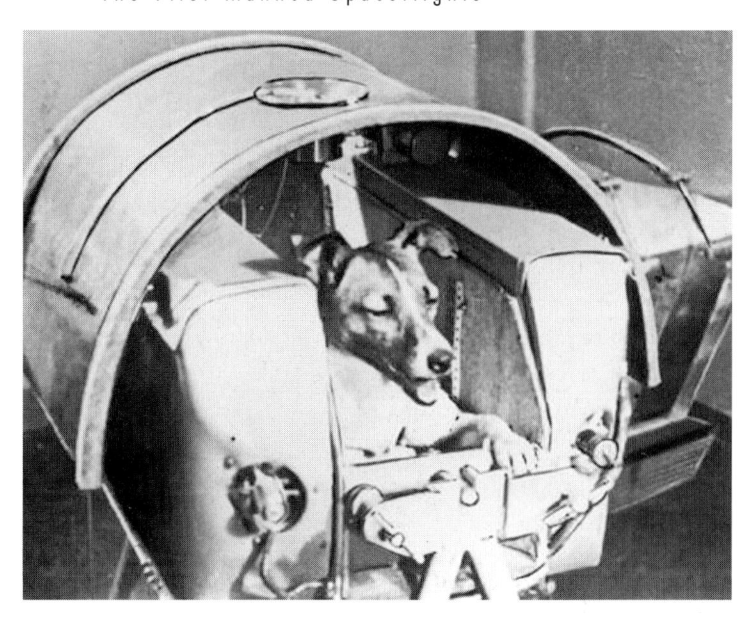

Laika in the capsule.

diameter of only 15cm. The satellite was built by the Jet Propulsion Laboratory (JPL) and was launched into orbit by a modified Jupiter-C (Juno 1) rocket from the Army Ballistic Missile Agency.

The satellite went into space and started to orbit the Earth at a perigee of 358km (222 miles) and an apogee of 2,250km (1,585 miles) with an orbital period of 114.8 minutes. The scientific instruments aboard the satellite consisted of the Iowa Cosmic Ray Instrument which discovered the radiation belts that circle the Earth. These became known as the Van Allen Radiation Belts after Dr James Van Allen who designed and built the detection instrument. The belt consists of charged particles trapped in the space surrounding the Earth by the Earth's magnetic field.

Explorer 1 stopped transmitting data on 23 May 1958 when its on-board batteries died, but remained in orbit for a further twelve years before re-entering the Earth's atmosphere over the Pacific Ocean on 31 March 1970. Between 1958 and 2007 90 Explorer satellites have been launched.

Explorer 1 was closely followed on 15 May 1958, when the Russians launched an unmanned space laboratory Sputnik 3. This was enormous in comparison to previous satellites and the cone, which housed the scientific instruments, was 11.7ft long and had a maximum diameter of 5.7ft. It weighed 2,919lb of which 2,129lb consisted of the scientific apparatus and power cells. Programmed to fly an elliptical orbit with a perigee of 226km and an apogee of 1,880km, Sputnik 3 carried out 10,037 orbits of the Earth.

On 29 July 1958 the National Air and Space Administration (NASA) was born, replacing the National Advisory Committee for Aeronautics (NACA).

In October 1958 that NASA became operational, taking over all the personnel and buildings.

On 13 December 1958 the first monkey was put into space. The squirrel monkey, named Gordo, was placed aboard a 50-ton Jupiter AM-13 rocket and launched 300 miles into space. The nose cone then detached and plummeted back to Earth, but a problem with the parachute caused the capsule to smash into the waters of the South Atlantic at 10,000mph and was lost. Both the Americans and the Russians continued to make progress towards the dream of putting a man on the Moon, but it was the Russians who were to make the first breakthrough.

The unmanned Russian spacecraft Luna I was launched on 2 January 1959. Two days later the spacecraft went into orbit around the Moon, and the battery-powered sphere orbited at altitudes between 5,000 and 6,000km and then, with its batteries depleted, went into a forty-five-day decaying orbit around the sun. The information sent back enforced the theory that there was no strong magnetic field around the Moon and gave additional information on the Earth's radiation belt. For the first time the strong flows of ionised plasma commonly known as the 'solar wind' were recorded.

On 28 May 1959 in America, two monkeys, Able and Baker, were launched into the atmosphere aboard a Redstone rocket with a Jupiter nose cone attached. Both monkeys were recovered after the flight, but Able died during an operation to remove the bio-electrodes from his head and body. Baker survived the operation and lived out his days at the Alabama Space and Rocket Center, Huntsville, Alabama. He died in 1984.

The next project was to put a man into space, and in America the Mercury project was started under the control of NASA. The design and development of the Mercury spacecraft was put out to tender and a number of contractors including Lockheed, Martin, Aeronautics, Avco, Goodyear, Convair, Bell, North American, Republic, McDonnell and Northrop submitted tenders. The one aircraft manufacturer that was missing was Grumman. The contract was awarded to McDonnell. The capsule had to be as small as possible to match the size of the Atlas rocket, which was the rocket chosen to launch the Mercury programme.

Seven astronauts were selected for training from the US military: three from the air force, three from the navy and one from the marine corps. They were Alan B. Shepard Jnr, USN; Virgil I. 'Gus' Grissom, USAF; L. Gordon Cooper, USAF; Walter M. Schirra Jnr, USN; Donald K. 'Deke' Slayton, USAF; John H. Glenn Jnr, USMC; and M. Scott Carpenter, USN. The 'Mercury Seven', as they were to become known, started training almost immediately. The criteria for the American astronaut-training programme were that the candidates were to be under 40 years old, less than 5ft 11in tall, be qualified test pilots, have a minimum overall flying time of 1,500 hours and, of course, be in excellent physical condition. The astronauts were put through exhaustive tests both physical and psychological. The psychological test started with a 600-question

The Mercury Seven astronauts selected for the Mercury space programme.

test on the personality alone, including twenty questions that started with the words: 'I am …'

The Soviet Union, too, had started to select cosmonauts. Soviet criteria differed from that of the Americans, inasmuch as the height of the cosmonauts was restricted to a maximum of 5ft 7in and a maximum weight of 154lb (11 stone). Their flying experience was extremely limited when compared with the Americans, i.e. Yuri Gagarin had only 240 flying hours, Alexi Leonov 250 hours and Gherman Titov 240 hours.

The cosmonaut tests included one in which the candidate had to spend several weeks in an isolation chamber, then was placed in a thermal chamber for a period of time, followed immediately by a similar period in a decompression chamber. On completion of these tests the candidate was placed aboard an aircraft and had to parachute out. The reasoning behind this test was to discover if the candidate could deal physically and psychologically with the rapid transition from being in a confined space to being released into boundless space in such a short period of time. All these tests were dangerous as they had never been tried before, but the results were to help set the yardstick by which all future cosmonauts would be selected.

A large number of Soviet pilots were initially chosen for selection, but this was quickly whittled down to six. The following were selected for further training:

Valentin Bondarenko.

Anatoli Kartashov; Valentin Varlamov; Yuri Gagarin; Gherman Titov; Andrian Nikolayev and Pavel Popovich. Two were dropped almost immediately due to accidents, Varlamov damaged his spine in a diving accident, whilst Kartashov: suffered a haemorrhage of the spine during training on the centrifugal force machine in which he experienced over 8G. A young 24-year-old trainee cosmonaut Valentin Bondarenko (L) took his place. After completing his initial tests satisfactorily, Bondarenko was placed in the isolation chamber with its rich oxygen atmosphere and had just finished his stint, when the atmosphere inside the chamber ignited into an inferno. It was discovered later that he had removed his bio-sensors and wiped his body with cotton wool and then thrown it carelessly onto an electric heater. The cotton wool caught fire and ignited the oxygen-rich atmosphere. Bondarenko died eight days later from severe burns. Like many of their accidents at the time, the Soviets kept it secret, fearful that the West may think they were having problems, and it was to be decades before the truth about this incident, together with a number of others, emerged.

One other cosmonaut was removed from the programme in 1961 after an alcohol-related incident. Gregori Nelyubov and two other trainee cosmonauts were returning from weekend leave when they got involved in a fight with a local army patrol. As long as they were prepared to apologise to the members of the army patrol then nothing more would be said about the incident. All the trainees with the exception of Nelyubov agreed. A report was filed to the cosmonaut's office regarding Nelyubov's arrogance and refusal to apologise. Nelyubov was dismissed from the programme and returned to a fighter unit. Depressed, he took solace in alcohol and, in 1966, committed suicide by walking in front of a train near Vladivostok. A photograph of the first seven cosmonauts showed Nelyubov, but a later photograph was released with him airbrushed out.

The launch site chosen by the Russians for their space programme was at Baikonur, in the middle of the southern region of Kazakhstan in the Soviet

Union. The area had been originally called Tyura but became known as Tyuratam by the nomads who travelled the region. Tam was the burial site of Tore-Baba, one of the descendents of Ghengis Khan, and its name was added to Tyura in his honour. The Russian authorities changed the name to Baikonur in an effort to confuse Western Intelligence as to the location of the rocket launch site. Baikonur was in reality a small town some 250 miles just east of the Aral Sea and to the north-east of the Syr Darya River. This was a pointless exercise, because the moment the first rocket (an R-7 ICBM) was launched from the site, an American radar station in Turkey picked it up and pinpointed the site. There was one other launch site, Kapustin Yar, situated 60 miles south-east of Volgograd, but this was used to launch the smaller type of rocket that was used for atmospheric research.

The R-7 motor was all right for the ICBMs, but to launch a rocket into space a much larger motor was required and this resulted in the creation of the R-16. The first test of this motor was disastrous. As the countdown approached launch

The Baikonur Cosmodrome.

time, it was discovered that there was a leak of nitric acid from the base of the unmanned rocket. Experienced rocket scientists would have carefully drained off the fuel and then pumped non-flammable nitrogen into the tanks to purge them. Instead, the Chief of Missile Deployment, Marshal Mitrofan Nedelin sent in engineers to try and fix the leak. In the firing blockhouse scientists should have reset all the electronic sequencers and then disarmed them, but they were ordered by Nedelin to delay the firing sequence and not cancel it. Somehow there was a confused signal given and the wrong command was transmitted to the rocket's upper stage and the motor fired. In the resulting explosion over 190 engineers, including Marshal Nedelin, were killed (the exact figure is not known) and an unknown number were badly injured.

Then, on 15 May 1960, the launch from Baikonur of Vostok 1K, the precursor to the manned space capsule, suddenly gave an air of urgency to the American space programme. The Russian spacecraft was placed in orbit around the Earth at a perigee of 284km and an apogee of 514km. The primary role of this spacecraft was that its main systems could be monitored over a long period to ensure its continuing safe flight and control. It finally went into a decaying orbit and burnt up on re-entry (it had no heatshield fitted) on 15 November 1965, after spending 1,979 days in space.

In July 1960 another test flight containing two dogs, Bars and Lisichka, was launched, but the rocket's booster exploded on the launch pad and both dogs were killed. Two further unmanned Sputniks were launched, Nos 3 and 4. Sputnik 3 was a geophysical satellite that explored the ionosphere and radiation belts around the Earth. It burnt up on re-entry after a lengthy period, but No 4 malfunctioned and it was to be five years before that too re-entered the Earth's atmosphere and burnt up. Then Korabl-Sputnik 5 was launched on 19 August 1960 with a veritable zoo aboard. Two dogs, Strelka and Belka, forty mice, two rats and fifteen flasks of fruit flies. After one orbit the capsule was separated and parachuted to the Earth. All the animals were safely recovered; the first time living things had been recovered safely from space. Sputnik 6 was launched on 1 December 1960 to measure the Earth's radiation belts amongst other things, but it burnt up on re-entry and is said to have had two dogs on board, Pchelka and Mushka. Both dogs died when the spacecraft came in at too steep an angle and was burnt up on re-entry.

Back in the United States, the first 'manned' American rocket flight into space was given to a chimpanzee called Ham. The name Ham was given to the chimpanzee after being raised and trained at the Holloman Aerospace Medical Facility. He was among three chimpanzees from the facility, and was selected after intensive medical and experimental tests. He was successfully launched on 31 January 1961 atop a Redstone rocket and had been chosen because, from a physiological standpoint, he closely resembled man and could be trained to carry out simple scientific experiments. Ham was taught to operate two levers,

The chimpanzee Ham in his contoured couch.

a red and a white one, in response to red and blue flashing lights. He was fitted with a pressure suit and strapped into a contoured couch, much the same as the astronauts that followed him. Just minutes after lift-off, the flight ran into problems when it was realised that the Mercury spacecraft had gone higher and faster than had been programmed. The mission was aborted and preparations made to bring the spacecraft back. Splashdown occurred much farther away than had been planned due to the craft's heightened speed, and it spent 2.5 hours in the water before being rescued by helicopter and taken to a recovery ship. During this period it took on 800lb of seawater and was in danger of sinking when recovered. Despite this, the mission was a success and Ham spent the rest of his life as a celebrity, first in the National Zoo in Washington and then later at the North Carolina Zoo. Ham passed away at the age of 26 in 1983 and is buried at the New Mexico Museum of Space History in Alamogordo. Ham's back-up chimpanzee for the flight, a female chimpanzee by the name of Minnie, never did get to fly.

On 9 March 1961, Korabl-Sputnik 9 was launched carrying a dog by the name of Chernushka. The spacecraft also carried a wooden dummy of a man, looking remarkably like Yuri Gagarin, made to the same dimensions and weight as one of the cosmonauts.

The spacecraft and the occupant carried out one orbit of the Earth before being recovered safely. On the 25 March Korabl-Sputnik 10 was launched with another dog on board by the name of Zvezdochka, and again was successfully recovered. Within four short years, the Russians had launched ten Sputnik spacecraft and three manned spacecraft – albeit all 'manned' by dogs.

Then at 0907 hours Moscow time, on 12 April 1961, came the breakthrough that was to change the face of space exploration, when Russian cosmonaut Yuri Gagarin was blasted into space from the launch site at Baikonur aboard Vostok I, call sign Kedr (Cedar), and became the first man to orbit the Earth. Gagarin circled the Earth for 1 hour and 48 minutes at a height of more than 300km and a speed of 28,000km/h.

During the flight Gagarin's controls were locked to prevent him taking over manual control, although in an emergency a key to unlock the controls was provided in a sealed envelope. There was one minor hiccup during re-entry when the Service Module, after being separated from the capsule, was found to be still connected by a wire loom. The spacecraft started to gyrate violently but fortunately the wire loom burned through. Gagarin ejected out of the spacecraft and parachuted to safety, landing near Engels Smelovka, Saratov. The Russians denied this for a number of years, only finally admitting it in 1978, after stating originally that the cosmonaut had landed with his spacecraft. Gagarin returned to Earth to a hero's welcome and this shy Russian air force officer's life was changed when his name became immortalised as the world's first spaceman.

The news that a man had been successfully launched into space and returned to Earth reverberated around the world. In the United States this was not what they wanted to hear and NASA officials quickly realised that they were falling behind the Russians. Their planned sub-orbital flight was rapidly brought forward and, on 5 May 1961, Commander Alan B. Shepard, USN, was blasted into a sub-orbit of the Earth in his spacecraft 'Freedom 7', built by the McDonnell Company, on top of a Mercury-Redstone III (MR-3) rocket. At one point during the launch, when delays were causing frustration amongst the Ground Controllers, Shepherd's calm voice came over the air:

> Why don't you fix your little problem and light this candle?

One of the problems, which held the launch for 52 minutes, was to replace a 115-volt, 400-cycle inverter in the electrical system on the launch vehicle. This was followed almost immediately by another problem: one of the Goddard IBM computers was found to contain an error. It needed a total re-check and configuration to resolve the problem, and by this time, coupled with weather delays, Shepard had spent 4 hours and 14 minutes strapped to the couch inside the tiny capsule. Eleven days earlier an attempt to place a mechanical astronaut into orbit had failed when the rocket was destroyed just 40 seconds after lift-off.

As the cables and the supporting boom fell away, Shepard started the elapsed-time clock. The lift-off started smoothly enough, but as the rocket passed through the transonic zone, the buffeting became violent and at one point Shepard's helmeted head was bouncing so hard he was unable to read the dials. Then suddenly the buffeting stopped and the transition to smooth flight passed through without incident. Shepard, now in constant voice communication with 'Deke' Slayton in the Mercury Control Center, informed Slayton that the dial-scanning procedure that had been planned would have to be dropped. Shepard decided that the need to watch his oxygen and hydrogen peroxide indicators was more important than watching the remainder of the dials.

Just after main engine cut-off, and 2 minutes and 32 seconds after launch, the tower-jettison rocket fired, followed immediately by the green tower-jettison light on the control panel becoming illuminated. Shepard then began the process of taking over manual control, one axis at a time. First the pitch control axis, which was controlled by means of a hand controller by his right hand. By moving the controller forward or backwards, Shepard was able to control the spacecraft's up and down movement. This was then followed by the pitch control, on the same hand controller, which controlled the left and right motion of the spacecraft by moving it left or right. At one point during the flight, Shepard moved his hand to remove a medium grey filter that he had inadvertently left over the periscope, and his hand banged into the abort handle. Very carefully Shepard pulled his hand away, holding his breath whilst he did so. Suffice to say, the filter stayed where it was, but the incident did nothing to detract from the breathtaking panorama that spread out beneath him.

As Shepard's 'Freedom 7' entered the Earth's atmosphere his rate of descent was much faster that he had expected. At 20,000ft the drogue 'chute snapped out, slowing the spacecraft fractionally, then 10,000ft later the main 'chute deployed and the spacecraft slowed reassuringly. Shepard then dumped the remaining hydrogen peroxide fuel and prepared for water impact. When it came, it was much softer that he had imagined.

Above the capsule, a US Marine Corps helicopter had watched the spacecraft make contact with the water and immediately moved into position directly above and hooked a cable through the recovery loops. Minutes later the astronaut and his spacecraft were on the deck of the aircraft carrier USS *Lake Champlain*. The flight lasted only 15 minutes and 22 seconds, and reached an altitude of 115.696 miles at a speed of 5,100mph, but it was the start for which the United States had been waiting. In a speech at a joint session of Congress on 25 May 1961, President John F. Kennedy all but challenged the Soviet Union by saying:

> I believe that this nation should commit itself to achieving the goal, before this decade is out, of landing a man on the Moon and returning him safely to Earth. No single space project in this period will be more impressive to mankind or more important in the long-range exploration of space; and none will be so difficult or expensive to accomplish.

The race was on!

Seven more sub-orbital flights were planned and the next, flown by Major Virgil 'Gus' Grissom, USAF, on 21 July 1961, was blasted into space aboard a Mercury-Redstone IV rocket (MR-4) in the spacecraft 'Liberty Bell 7'. The pre-launch activities were running normally but suddenly there was a call to halt the countdown when one of the seventy gantry bolts was found to be misaligned. It was decided that the remaining sixty-nine bolts were more than sufficient to hold and, if required, blow the hatch. Grissom, alone in the tiny capsule on top of the Redstone rocket, was being monitored constantly and it was noticed that when the countdown hold was activated, his pulse rate soared from 64 to 162 pulses per minute. The moment the main engine fired, Grissom relaxed and settled back to enjoy the ride. After 2 minutes and 22 seconds the main engine shut down and it was at this point the craft entered space. The capsule tumbled momentarily whilst the turnaround was automatically carried out, then manual control was passed to the pilot. For the first time since the launch, Grissom was able to see some sort of reference and waxed lyrical about the 800-mile arc of the Earth's horizon. Grissom then carried out a number of pitch and yaw movements by means of the hand controller, then carried out a roll-over manoeuvre so that he could see the east coast of the United States as he sped over. Unfortunately he got so carried away with the sightseeing that he fell way behind in his work commitment.

Suddenly it was time to return to Earth and the spacecraft was put into a re-entry position of 14 degrees from the Earth vertical. The re-entry was text-book until splashdown in the sea. Having established contact with the recovery helicopter's pilot, Lieutenant James L. Lewis, Grissom relaxed in his couch and unbuckled his harness. Suddenly the capsule's hatch blew open prematurely and water began to pour in. Grissom realised that there was nothing that he could do and abandoned the space capsule.

Lewis approached the sinking spacecraft, leaving the second helicopter to pick up Grissom. The crew member on board the helicopter picked up the 'shepherd's hook' recovery pole and carefully threaded the crook through the recovery loop on top of the spacecraft. By this time Lewis had lowered the helicopter to a point that the chopper's three wheels were almost in the water. Liberty Bell 7 sank out of sight, but the pickup pole twanged as the attached cable went taut, indicating to the helicopter pilots that they had made their catch. But at that moment Lewis called a warning that a detector light had flashed on the instrument panel, meaning that metal chips were in the oil sump because of engine strain. The pilot,

Mercury-Redstone IV lifting off the launch pad.

Marine helicopter attempting to lift Grissom's capsule from the water.

watching his insistent red warning light, decided not to chance losing two craft in one day. As the first helicopter moved away from Grissom, it struggled to raise the spacecraft high enough to drain the water from the impact bag. Once the spacecraft was almost clear of the water, but like an anchor it prevented the helicopter from moving forward. The flooded Liberty Bell 7 weighed over 5,000lb (2,300kg), 1,000lb (450kg) beyond the helicopter's lifting capacity. He finally cast loose, allowing the spacecraft to sink swiftly.

Meanwhile, Grissom realised that he was not riding as high in the water as he had been because he had forgotten to secure his suit inlet valve. Swimming was becoming difficult, and now with the second helicopter moving in he found the rotor wash between the two aircraft was making swimming even harder. Bobbing under the waves, Grissom was scared, angry and looking for a swimmer from one of the helicopters to help him tread water. Then he caught sight of a familiar face, that of George Cox, aboard the second helicopter. Cox was the co-pilot who had retrieved both the chimpanzee Ham and Shepard on the first Mercury flight. With his head barely above water, Grissom found the sight of Cox heartening.

Cox tossed the 'horse-collar' lifeline straight to Grissom, who immediately wrapped himself into the sling backwards. Lack of orthodoxy mattered little to

Grissom now, for he was on his way to the safety of the helicopter, even though swells dunked him twice more before he got aboard. His first thought was to get a life preserver on. Grissom had been either swimming or floating for a period of only 4 or 5 minutes, 'although it seemed like an eternity', as he said afterwards.

Astronaut Walter Schirra was assigned to carry out tests on the plunger that activated the hatch, and after a series of in-depth examinations and practical tests he came to the conclusion that they would never find out what really happened. But one thing he did say was that, in his considered opinion, 'There was only a very remote possibility that the plunger could have been actuated inadvertently by the pilot.' The Board of Inquiry cleared Grissom of any misdoing and he was exonerated. The capsule wasn't recovered until 1999.

Two weeks later, on 6 August 1961, the Russians launched their second manned spacecraft, Vostok 2, call sign Oryel (Eagle). The cosmonaut was Major Gherman Stepanovich Titov, who had been Gagarin's back up on the first manned flight. Titov, at the age of 25, was the youngest person to go into space and probably still is. He carried out a 17-orbit, 25-hour flight, which was way beyond anything that the Americans had been able to even dream of doing.

Titov's main mission was to find out how a human being would bear up under excessive 'G' loads and prolonged weightlessness whilst carrying out tasks. The programme was that the spacecraft would carry out seventeen orbits of the Earth during the day, each orbit carrying different work schedules. After the sixth orbit he was given authorisation to carry out manual control of the spacecraft. This was a crucial moment as it was the first time a cosmonaut had complete control of his spacecraft whilst in space. The spacecraft responded to his touch immediately and with great relief he radioed the news back to base. Titov carried a number of manoeuvres during the following orbits, and took numerous photographs of the Earth together with some film footage. He also took time out to eat and drink which was an important part of his programme because of the need to provide future cosmonauts with nourishment on longer spaceflights.

Titov suffered from space sickness during the mission which hampered some of the experiments he had been assigned to carry out. After the seventeenth orbit, Titov's radio crackled into life and the order came for him to prepare for re-entry. During re-entry he experienced a similar problem to that of Gagarin: the separation of the Service Module from the capsule. Fortunately the problem resolved itself and, given the choice of staying with the capsule or ejecting, Titov chose to eject from the spacecraft and landed in a field near Krasny Kut, Saratovskaya Oblast.

Gherman Titov never flew again; he was assigned to carry out testing and research on the Spiral spaceplane, which many had declared to be a dead-end project before it had even started. Titov died of a heart attack in September 2000.

On 20 February 1962 Lieutenant-Colonel John Glenn, USMC, was launched aboard his Mercury spacecraft MA (Mercury-Atlas) VI, call sign Friendship 7. The day had started at 0220 hours that morning, when John Glenn was woken. By

John Glenn beside his Friendship 7 spacecraft.

0500 hours he was in his spacesuit and in a van on the way to the launch pad. The countdown had started, but already delays were starting to creep in; the replacement of some of the electronic guidance system and the weather forecast was not looking too good. At 0603 hours John Glenn was assisted into the spacecraft and as he was being eased into his contoured couch it was noticed that the respiration sensor, which was a thermistor attached to his microphone, had slipped. The only way it could be put back into position was by opening up the suit. This was something that could not be done safely on top of the launching gantry, so it was decided to leave it where it was. Mercury Flight Control was then told to ignore all respiratory transmissions, as they would probably be wrong.

As the technicians started to bolt the seventy bolts of the hatch onto the spacecraft, a broken one was discovered. Gus Grissom had flown his mission with a broken hatch bolt, but Walter Williams, NASA's chief engineer, decided that this was not right and he ordered it to be removed and repaired. This put the launch on hold for a further 40 minutes, so the engineers decided to run a further check on the guidance system, part of which had been replaced earlier. With all this done and the hatch bolted on, a further hold was announced to add more propellant to the tanks on the booster. With the booster tanks full, there came yet another hold - a stuck fuel pump outlet valve. Finally, at 0947 hours, and 3 hours and 44 minutes after Glenn had slid into the capsule, the Atlas rocket, with the Friendship 7 spacecraft perched precariously on top, lifted off the pad.

As the spacecraft passed over the tracking station at Guaymas, telemetry signals indicated that there was a problem with one of the yaw reaction jets, which would give Glenn a problem with his attitude control. These suddenly reminded the engineers at Mercury Flight Control of the last unmanned flight, MA-5, when a similar problem had caused the spacecraft to be brought back earlier than planned. Glenn was made aware of the problem and switched from the automatic stabilisation and control system to manual-proportional control mode. Then, by switching between modes, he found the most effective means by which to control the spacecraft. Although the yaw reaction jet came back on line, Glenn chose to fly the spacecraft manually, proving beyond all doubt that man was necessary for spaceflights because they could augment the reliability of the machine.

Then a telemetry signal from the spacecraft indicated that the heatshield and the compressed landing bag had come loose. This became a major concern because they both were only held in place by the straps of the retropackage. If the retropackage was jettisoned after re-entry, then the heatshield would come off. If left on, then any unburned solid propellant would ignite. It was decided that the retropackage would not be jettisoned and the dynamic forces of the spacecraft would keep the heatshield in the correct position. Glenn was asked to double-check the landing bag deploy switch, which made him a little wary that something was not quite right. Flight control had decided not to tell Glenn of the problem as they thought he had enough to concern him. But it became obvious when a message from the Hawaiian tracking station told Glenn to put the landing bag deploy switch into the automatic position and if the light should come on, he should enter the atmosphere with the retropack in place.

As the spacecraft entered the Earth's atmosphere, Mercury Mission Control, tracking stations around the world and engineers on the Mercury project held their breath. Glenn, in the capsule, was experiencing severe oscillations so he switched control of the spacecraft to manual. His remark to flight control about 'That's a real fireball outside' did nothing to dispel the fears of the men on the ground. Then suddenly at 28,000ft the drogue 'chute snapped out and all the oscillations ceased. At 17,000ft the main 'chute deployed and Friendship 7 drifted gently into the sea. Twenty minutes later the destroyer USS *Noa* came alongside the bobbing spacecraft, a line was attached and it was hauled up onto the deck. It transpired later that a malfunctioning switch inside the spacecraft caused the heatshield problem.

The flight lasted 4 hours and 55 minutes and could not be fully compared with Gherman Titov's flight, but American pride had been partially restored and they were back in the race.

Friendship 7 being recovered from the sea.

The fourth Mercury-Atlas flight, MA-7, call sign Aurora 7, could have ended in disaster. The astronaut selected for the flight was Captain Donald K. 'Deke' Slayton, USAF, but it was discovered during a routine medical that he had a slight heart murmur. Such were the stringent requirements regarding health and the physical condition of astronauts that his place was taken almost immediately by Commander Scott Carpenter, USN.

Despite three 15-minute holds, the launch on 24 May 1962 was the smoothest one to date and was watched on television by an estimated 40 million people. At 38,000ft the escape tower jettisoned and Carpenter watched it cartwheel away, as if in slow motion, toward the horizon, with smoke still trailing from its three rocket nozzles. With the spacecraft in orbit, Carpenter, like Glenn, commented on the fact that he felt no sensation of speed, despite the fact that he was now travelling at 17,549mph.

It is interesting to note that the food taken by Carpenter on his flight differed greatly to that taken by John Glenn. Glenn had taken squeeze tubes of the baby-food type, whereas Carpenter had had a variety of foods prepared by the Pillsbury Company that consisted of chocolate, figs and dates. The Nestlé Company had made some bon-bons composed of high-protein cereals, orange peel with almonds and cereals with raisins. All these foods were covered with an edible glaze. It soon became obvious that in a weightless environment the squeeze tube variety was the most practical, as when Carpenter bit into one of the bon-bons, the crumbs floated about the cabin. This may sound innocuous, but it could have been potentially lethal if any of the crumbs had got behind the electrical panels and caused a short circuit. It was also difficult to eat with gloved hands.

Things then started to go wrong with the flight, partly due to problems outside of Scott Carpenter's control and partly due to his over-eagerness. Taking over manual control of the spacecraft, Carpenter experimentally rolled and yawed the capsule in every way he could, including at one point standing it on the antenna canister end. This was in addition to carrying out the designated experiments that were part of his mission brief. Then he inadvertently activated the sensitive-to-the-touch, high-thrust attitude control jets. This caused both the automatic and manual systems to become temporarily redundant. Scott Carpenter was oblivious to the fact that he was using up fuel at an alarming rate of knots, and flight control continually had to remind him of the danger.

When the time came for him to hit the retrofire button to start re-entry procedures, the spacecraft's attitude was about 25 degrees to the right, which would cause an overshoot of about 175 miles. Coupled with the fact that the retrorockets began firing 3 seconds late, and that the thrust from the rockets themselves was not as powerful as it should have been, an additional 60 miles was added to the overshoot. As the rockets fired, Carpenter realised that he was still on manual control, but the fuel gauges were now reading empty on manual and only 15 per cent on automatic. Then he was into re-entry and everything was now academic. The communications blackspot began and all transmissions ceased. Carpenter felt very alone, and the spacecraft started to oscillate wildly. At 25,000ft he released the drogue 'chute, and as it deployed the spacecraft slowed fractionally. The main 'chute was armed at 15,000ft and deployed manually at 9,000ft. Then came the comforting sound of the parachute snapping open followed, minutes later, by the reassuring jolt of the spacecraft hitting the water – 125 miles off the planned landing target.

The spacecraft lay in the water at an angle and Carpenter decided to wait to see if it would stabilise. After a few minutes he realised that this was not going to happen and that the spacecraft was starting to lie quite deep in the water. Alan Shepard had told him just prior to landing in the water that because of his overshoot it would be at least an hour before an aircraft could get to him. Carpenter decided that it was too risky to blow the hatch and so decided to exit the spacecraft through the neck. After a great deal of difficulty, he struggled through the top of the spacecraft and inflated a yellow life raft. Clambering into it, he waited for the rescuers to arrive, well aware that the raft carried no radio. Fortunately, although there was a heavy swell, the weather was not inclement. After about half an hour there came the drone of an aircraft's engines followed closely by a P2V Neptune and, of all things, a small Piper Apache. It soon became obvious that the Apache was just there to take photographs. Then after 20 minutes, a SC-54 aircraft arrived and out dropped two frogmen, who quickly swam to Carpenter in his life raft. They swiftly inflated two other life rafts that they had brought with them and locked the three together.

After over 3 hours of sitting in his life raft, Carpenter and the two frogmen were picked up by a helicopter from the aircraft carrier USS *Intrepid* and taken back to the ship, none the worse for their ordeal. There were recriminations about the delay in picking up Scott Carpenter, most of them levelled at the astronaut himself. The overshoot was down to him and he was the first to admit this. But to be fair to Scott Carpenter, he had had less than ten weeks to prepare for his flight and this was a major contribution to his problems. Fortunately he and the spacecraft were recovered successfully, but because of the criticism levelled against him by the NASA hierarchy, he was never selected to fly in space again. Instead he turned his talents to becoming an aquanaut in the US navy's 'Man of the Sea' programme.

In Russia preparations were under way for another spaceflight, Vostok 3, whose call sign Sokol (Falcon) was to be flown by cosmonaut Andrian Nikolayev. The flight was supposed to have taken place in March, but because an explosion during the launching of a Zenit-2 reconnaissance satellite had damaged one of the two Vostok launch pads, the launch was put back until 11 August. Vostok 3's mission initially was to rendezvous with Vostok 4, which was to be launched the following day. The launch from Baikonur went according to plan and Vostok 3 slid into orbit at an apogee of 218km above the Earth.

Nikolayev settled down to wait for fellow cosmonaut Pavel Popovich in Vostok 4, call sign Berkut (Golden Eagle), to be launched the following day, but in the meantime he had a number of experiments and scientific observations to carry out. Further tests were done regarding the cosmonauts' ability to function under weightless conditions and to undertake certain experiments. The main objective for this particular mission was to carry out a rendezvous with another spacecraft. Neither spacecraft had any real manoeuvring capability so a link-up was not possible, but it did give the ground Mission Controllers practice in handling two spacecraft in such close proximity. After a quiet night, Pavel Popovich's spacecraft was manoeuvred to within 6.5km of Nikolayev in Vostok 3, and the two spacecraft carried out their manoeuvring exercises. During transmission with Gagarin, who was acting as CapCom (Capsule Communicator), Nikolayev looked out of his viewport and saw Vostok 4 in the distance.

All the while the mission and the cosmonauts were being televised and transmitted to the people in Russia. Pavel Popovich became the world's first space television commentator when he gave a running commentary on his observations of the Earth as seen from space. After completing his entire mission Nikolayev returned to Earth on 15 August, close to where Pavel Popovich had landed. Both spacecraft had been scheduled for a four-day mission, but Popovich was brought down a day early because of a misunderstanding.

The misunderstanding came about because of the extreme paranoia that afflicted the Russian space scientists and political hierarchy. Believing that everything that was said between the cosmonauts and the Flight Controllers was being monitored by the West, they dreamt up codewords and expressions to pass sensitive messages back and forth between the cosmonauts and their controllers. In the case of Pavel Popovich, he had been instructed that if he felt ill due to space sickness, he was to say that he was 'observing thunderstorms'. As his spacecraft flew over the Gulf of Mexico, Popovich actually observed a most violent thunderstorm and transmitted this information back to his controller at Baikonur. He was recalled immediately and when back on Earth it was discovered that there was nothing wrong with him and he had been more than willing to continue the flight.

In the United States the Mercury programme was coming to an end after two more flights, MA-8 (Walter Schirra) and MA-9 (Gordon Cooper). A proposal was put forward that a programme that would carry two men into space should replace it. A NASA employee by the name of Alex P. Nagy had chosen the name Gemini for the new programme. It was decided that two planned Mercury flights would go ahead, but the proposal put forward to Congress in April 1959 was for funding in the sum of $3 million to enable research into space rendezvous techniques to be carried out. This would include the design and research of a manned space laboratory and a two-man spacecraft.

In the meantime, Mercury VIII, MA-8, call sign Sigma 7 was on the pad at Cape Canaveral waiting for astronaut Walter 'Wally' Schirra to climb aboard. On 3 October 1962, Schirra slipped into the contoured seat in the Mercury capsule and started to carry out all his pre-flight checks. Whilst in the process of stowing away all his bits and pieces ready for the flight – star charts, camera and the like – he found a steak sandwich that had been smuggled into one of the compartments by a fellow astronaut. Ruefully smiling to himself, and thinking of the tubes of tasteless food that had been assigned to the programme, he was glad that he would have something substantial to eat whilst in space.

As with all the other Mercury flights, the countdown started well enough, but then came the 'hold' signal. One of the radar sets at the tracking station in the Canary Islands had malfunctioned and it was a necessary part of the mission that this particular one was up and running. Fortunately it only took 15 minutes to fix the problem and the countdown was restarted. At 0715 hours the pad rumbled under the power of the Atlas rocket and then Mercury VIII was climbing slowly upwards and heading into space. Then 10 seconds later the rocket developed an unexpected clockwise roll. Seconds later the roll stopped, much to the relief of everybody in the flight control, and the rocket continued on upwards.

Schirra, well aware of what was happening, had his hand on the abort handle and, had the roll not sorted itself out, he would have pulled it. The booster engine kicked in and pushed the spacecraft onward and upward, then cut out. Seconds later the escape tower separated as its rockets ignited. The sustainer engine continued and then Sigma 7 settled into orbit. With his eyes glued to the instrument panel, Schirra transferred control to the fly-by-wire mode and started to obtain his correct orbital attitude position. Realising the problems that had risen regarding the excessive use of fuel during Scott Carpenter's flight, Schirra ensured that all his control movements of the spacecraft were slow and deliberate. With the spacecraft now in a stabilised orbit, Schirra allowed himself the luxury of a few minutes of sightseeing as he passed over the African continent.

Schirra placed control of the spacecraft on automatic and started to carry out some of the experiments that had been built into the mission. As the night sky rushed to meet him, he aligned the periscope towards it, but like the astronauts before him, found it to be a 'useless piece of baggage'. One of his tasks was to carry out a series of alignments when the spacecraft yawed from the flight path. He also carried altitude correction from a visual aspect, using the reticule in the window. During the night pass over Australia on his second orbit, Schirra carried out one of the hardest experiments of the flight. By using celestial navigation in conjunction with the star-finder charts, he was to carry out an alignment of the spacecraft by positioning it in relation to known stars or planets, and then by watching the apparent motion of the stars he would test his sense of facing left or right. The experiment worked successfully, but was extremely time-consuming and tiring. Nevertheless, it was an experiment that was very necessary, as by its very nature it would be an exceptionally important skill to have in the event of an emergency.

As he approached his final orbit, Schirra started to stow everything that was not required and go through his pre-entry checklist. Taking the spacecraft off automatic mode and shifting it to fly-by-wire, Schirra prepared it for retrofire. At the right moment he fired the retrorockets and the spacecraft slowed markedly. As he entered the atmosphere, Schirra watched the altimeter and as it reached the 40,000ft mark he punched the drogue 'chute button. As it deployed, he activated the jettison switch to get rid of the excess fuel. At the 15,000ft mark he manually deployed the main 'chute and watched with great relief as it streamed out, unfurled and then snapped open.

Gordon Cooper being helped into his spacecraft.

Gordon Cooper in his spacecraft just prior to lift off.

The spacecraft gently splashed down just as swimmers and helicopters from the aircraft carrier USS *Kearsarge* arrived. Within half an hour the spacecraft, with Schirra still inside, was lifted onto the deck of the carrier. The flight was textbook and the information gathered from the experiments was worth its weight in gold.

The last of the Mercury flights, MA-9, was on 15 May 1963 and was flown by astronaut Gordon Cooper. The launch went as planned without the usual expected delays and at 0804 EDT, Faith 7 lifted off the pad and climbed skywards. Two minutes and 14 seconds later, Gordon Cooper heard a sharp thud as the booster engines cut off. The escape tower blasting clear of the capsule closely followed this. The sustainer engines continued to push the rocket toward the atmosphere, the guidance system controlling the flight for the 2 minutes before the Sustainer Engine Cut Off (SECO) sequence was activated. Once in orbit Gordon Cooper switched control from automatic to fly-by-wire and hurtled over Bermuda at 17,547mph. Before he knew it, tracking stations were calling him one after another whilst he worked frantically to confirm the telemetry information requested.

On the second orbit Mercury Flight Control informed Cooper that his orbital parameters looked so good and that all the telemetry signals from his spacecraft were functioning so well that they had scheduled his flight for a possible twenty orbits. Cooper had discovered that his suit temperature was fluctuating erratically, but not enough to cause him any major concern. For a while he forgot everything as he marvelled at the unblinking stars in space and the different weather systems moving around the Earth.

When the third orbit came around, Cooper started to check on some of the experiments for which he had been scheduled, eleven in all. One was to eject a

6in diameter sphere fitted with polar exon strobe lights. The idea was that on his next pass he would be able to test his ability to spot and track this flashing beacon whilst in orbit. After launching the beacon he scanned space on his fourth orbit, but he could not see anything remotely like a flashing signal, but on the fifth pass there it was! With unreserved delight, Cooper reported, 'I was with the little rascal all night'.

Cooper continued to keep check on all his biological functions: blood pressure, temperature, urinary samples and even calibrated exercise. As he passed through his seventh orbit, he was in the process of transferring the urinary samples from one tank to another when he was made aware that he had passed Wally Schirra's record of orbits. He was also having problems pumping the samples from one tank to the other. He placed on record: 'The thing about this pumping under zero g is not good. [Liquid] tends to stand in the pipes and you have to actually force it through.'

The flight continued as planned and by orbit fifteen the only problem Cooper was having was that the heat exchanger in his suit kept fluctuating between hot and cold. Orbits sixteen, seventeen and eighteen passed without incident, just more experiments, checking and photographs. Then, during the nineteenth orbit, a small green light appeared on the instrument panel, indicating that the spacecraft was decelerating, which could affect the altitude stabilisation at the point of retrofire. The problem was further exacerbated on the next pass when all the attitude readings were lost. Mercury Flight Control was showing deep concern, but this was nothing compared to the concern they showed when on the next pass, the twenty-first orbit, a short circuit occurred in a busbar that served the 250-volt inverter. This left the whole of the stabilisation and control system without electric power. There really was now a very serious problem.

The time had come to bring the spacecraft home. Gordon Cooper remained remarkably calm throughout the problem and, by the time he had completed his checklist with John Glenn, all was ready to bring the spacecraft back under manual control. The re-entry flight and the deploying of the parachutes went perfectly and the spacecraft landed in the sea just 4 miles from the aircraft carrier USS *Kearsarge*. Within minutes a helicopter was hovering overhead and the capsule was winched out of the water. Forty minutes later the spacecraft, with Gordon Cooper still inside, was deposited on the deck of the ship. After blowing the hatch, doctors examined Gordon whilst he was still in the spacecraft to make sure that he was all right. Despite all the accolades that came his way, Gordon Cooper kept his feet firmly on the ground; he knew that over the next few months his life was going to be one round of celebrity appearances intermingled with in-depth technical debriefings.

No more Mercury flights were planned; NASA now had sufficient information and knowledge to put into action the next phase of their space programme – Gemini.

On the other side of the world, Russia launched Vostok 5 with a call sign of Yastreb (Hawk), from Baikonur, with cosmonaut Valery Fyodorovich Bykovsky at the controls on 14 June 1963. The purpose of the flight was to extend the information of the various spaceflight factors obtained on previous Vostok flights. The mission was scheduled for an eight-day flight, but it was curtailed at the onset due to technical problems. Due to further technical problems after launch, Vostok 5 was placed into a lower orbit than was planned and was noticeably going into a decaying orbit every time it went round the Earth. Then, after only four days, Bykovsky contacted Mission Control saying that he had severe problems with his waste management system and that conditions in the cramped capsule were becoming extremely uncomfortable. The controllers realised that nothing more was to be gained from the flight and ordered it back to Earth. As the spacecraft entered the atmosphere, the Service Module was jettisoned, but failed to release completely causing the capsule to spin violently. Fortunately the heat from the re-entry burned off the offending retaining strap and the spacecraft returned to its normal re-entry mode and effected a safe home landing for Bykovsky.

Romance featured in the space race when, in June 1963, cosmonaut Andrian Nikolayev's girlfriend, 26-year-old Valentina Vladimirovna Tereshkova, joined the Soviet space programme as a cosmonaut. Her first flight on 16 June 1963 was aboard Vostok 6 and was to have been made as a dual flight with cosmonaut Fydorovich Bykovsky who was already in orbit aboard Vostok 5.

Two days after the launch of Vostok 5, the last of the Vostok programme, Vostok 6, was launched. Aboard the spacecraft, call sign Chayka (Seagull) was Valentina Tereshkova, Russia's first female cosmonaut. Because of the problems experienced by Bykovsky, it was decided to just use the flight to carry out research on the effect and comparisons of spaceflight on male and female organisms. It was mooted at one point that putting a woman into space at the time was more of a novelty than a serious mission. This thought seems to have been endorsed when Sergei Korolev, the director of space programmes, was unhappy with Tereshkova's work rate and performance whilst in orbit and would not allow her to take over manual control of the spacecraft. Suffice to say that she never flew again. Tereshkova's flight lasted two days, in which she made forty-eight orbits of the Earth and carried out an extensive flight programme. Bykovsky, on the other hand, spent five days in space, completed eighty-one orbits of the Earth and carried out numerous physical and psychological tests. Valentina Tereshkova later married fellow cosmonaut Andrian Nikolayev.

The recovery of these two flights marked the end of the Vostok programme and heralded the arrival of a new spacecraft, the Voskhod. This type of rocket was transported to the launch site on a purpose-built railroad carriage and track. The new spacecraft was much larger than its predecessor inasmuch as the pressurised cabin area could accommodate three cosmonauts and was designed so that the crew could operate in a shirtsleeve environment without the need for spacesuits and helmets.

Valentina Tereshkova just prior to boarding her spacecraft.

It was equipped with a reserve retrorocket engine, improved communications and telemetry equipment, and a second orientation system that used ionic sensors. The spacecraft could be controlled manually or automatically both whilst in flight and when landing. A series of unmanned flights were carried out until Korolev was satisfied that the soft landing recovery system worked perfectly.

But NASA had not been idle. Since the beginning of the 1960s it had been carrying out experiments with 'lifting bodies'. These were small, wingless aircraft, some equipped with a rocket motor, that were dropped from a converted Boeing B-52 bomber and some of which were merely gliders. The object of the exercise was to see if a re-usable spacecraft could be developed, one that could manoeuvre in space and then land back on Earth. The idea of the lifting bodies was that the aerodynamic shape of the fuselage would act as wings as the aircraft re-entered the atmosphere.

Martin Marietta X-24A.

Boeing was the first contractor to produce a design for USAF/NASA. Their concept, in 1959, was for a piloted, manoeuvrable vehicle that could operate in the hypersonic and orbital flight areas. It was called the X-20 or Dyna-Soar (Dynamic-Soaring). None were ever built as the programme was cancelled, although a number of mock-ups were made to show the basic design. The programme was re-initialised in 1962 when the Martin Marietta Company produced the X-24A (also known as SV-5P). Only one was ever built, and after being dropped from beneath the wing of a converted Boeing bomber NB-52B, was flown successfully.

Then came the design known as the M2 (NASA 817). Again this was one of the blunt-body designs, and a strange looking aircraft, as indeed nearly all the lifting bodies were. The M2 was ostensibly a glider fitted with a solid-fuel rocket motor in case it was necessary to assist the aircraft in a pre-landing flare manoeuvre. The initial flight of the M2F1 was planned using a van to tow the little aircraft along the runway until it had enough speed to get lift. But not one of the vans or trucks was fast enough. Then someone came up with the bizarre idea of using a stripped down Pontiac convertible, with a four-barrel carburetor and a manual gearbox that did 8 miles to the gallon. The car was handed over to a custom-car dealer in Long Beach, California, who fitted roll bars, special rear-facing observer seats and radio equipment. The car was then sprayed the statutory high-visibility yellow of all flight-line vehicles and had 'National Air and Space Administration' painted on the sides. Looking more like a refugee from a hot rod meeting than a serious piece of research equipment belonging to the space administration, the lifting body tests got under way.

The later version of the M2F1, the M2F2, was designed to be dropped from beneath the wing of the converted B-52B bomber that had previously been used for the X-24A tests.

The Americans were thinking way ahead on the space programme as, before they had even launched their first two-man spacecraft they were carrying out tests on a three-man spacecraft – Apollo. Abort tests were being carried out at White Sands Missile Range, New Mexico, in December 1963, on the Apollo spacecraft in the event of there being a problem during the launch.

On 10 December 1963 Secretary of Defence Robert McNamara announced the start of the Manned Orbiting Laboratory (MOL) project and the cancellation of the Dyna-Soar programme. The MOL was actually called Gemini B/MOL and was originally the brainchild of the USAF. The idea was to place a Gemini spacecraft on top of a cylindrical laboratory, then place the pair on top of a Titan II rocket and launch the two as one unit. Having reached orbit, the MOL would continue to orbit the Earth whilst the Gemini spacecraft would be used as a shuttle between the MOL and the Earth. The proposed recovery of the Gemini spacecraft was quite unique. Instead of it splashing down in the sea on parachutes,

M2F1 being towed in flight.

M2F2 beneath the wing of a converted B-52B bomber.

as with the Mercury and Gemini programmes, it was proposed that a paraglider wing would be deployed and the spacecraft would land on the ground, much like today's paragliders.

Two launch sites were selected, Cape Canaveral and Vandenberg, but it was to be four years before the first real moves to build the MOL were made. In February 1967, Douglas was named as the prime contractor to build the orbiting laboratory. With the building of the MOL came problems. As the orbiting laboratory developed so did the weight and the USAF had to consider upgrading the engines that were to place the MOL in orbit. The cost was also spiraling from a projected cost of $1.5 billion to $2.2 billion and the first flight was re-scheduled for 1970. In order to save money the Vandenberg site was cancelled as a launch site and all work was concentrated on Cape Canaveral.

CHAPTER THREE

THE FIRST MULTI-CREWED SPACEFLIGHTS

The Soviet Union was now struggling to keep pace with the Americans, after having led the race for so long. On 12 October 1964 they had launched the first multicrew spacecraft, Voskhod 1, from their launch site at Baikonur. The crew, Vladimir Komarov, Konstantin Feoktistov and Boris Yegorov had earlier in the year seen the successful launch of two unmanned Voskhod type spacecraft. The three cosmonauts wore no spacesuits during the launch or whilst in flight, as it was thought that the spacecraft were now so well pressurised that they were unnecessary. This was to later prove fatal during the Soyuz 11 mission.

Voskhod 1 entered orbit at a perigee of 178km, reaching an apogee of 336km. Although the spacecraft was now designed to accommodate three cosmonauts, the designers had given very little thought to the re-siting of the instrument panel. They left the panel in exactly the same place as it was in the Vostok spacecraft. The crew completed a number of experiments before returning to Earth the following day. They had planned for a longer flight, but back on Earth the Russian President, Nikita Khrushchev, had been removed from power whilst they were in orbit and they were ordered back to Earth by flight control.

Two days before the Americans were to launch their first Gemini flight, on 23 March 1965, the second of the Voskhod missions, Voskhod 2, was launched from Baikonur. This time there were only two cosmonauts on board, Alexi Leonov and Pavel Belyayev, and they were to carry out the first Russian spacewalk. The mission was fraught with problems from the first. After clambering into his spacesuit, Alexi Leonov carried out his spacewalk, but on his return he had the greatest difficulty in getting back inside the spacecraft due to the stiffness of his suit and at one point it was thought that he wouldn't get back at all. He had to release most of the air from his suit to regain some semblance of manoeuvrability before struggling back inside the spacecraft. In the event that he was unable to get back and Belyayev had to leave him in space, Leonov had a suicide pill secreted in his helmet for just such a contingency.

Voskhod 1 on the launch pad.

Once back inside the spacecraft the problems didn't end there as, after getting Leonov back in, it was found that the primary hatch would not seal properly. To compensate for this the environmental control system was activated and the capsule was flooded with pure oxygen instead of the nitrogen-oxygen mixture that normally was in the spacecraft. This, of course, was a potential fire hazard, as was later discovered in 1967 when the three Apollo I astronauts died whilst in a similar environment.

After the spacewalk the craft prepared for re-entry, but the primary retrorockets failed to fire so the spacecraft carried out another orbit of the Earth. At the point of re-entry this time the retrorockets were fired manually. On entering the Earth's atmosphere the Service Module failed to release completely, causing the capsule to gyrate violently before the wires holding it burned through. The spacecraft landed deep inside a forest near Perm in the Ural Mountains, 2,000km away from the nearest rescue teams. The crew had to spend the night in their capsule, surrounded by wolves, before being rescued the following day. Soviet doctors later reported that Leonov almost suffered heatstroke whilst in space, a condition that could have been fatal in that environment. Despite being fraught with problems, the mission was a success.

The Americans, however, were also continuing to make progress with the Gemini project which had come to the fore in April 1959. Originally named Mercury Mk.II, the Gemini spacecraft built by McDonnell was the first with the ability to change orbit, have a rendezvous and docking facility, enable the astronauts to leave their environment and return, and carry out long duration spaceflights. The launch vehicle chosen to blast this spacecraft into space was a modified US Air Force Titan II ICBM. Three metres in diameter and 33.2m tall, the two-stage rocket was fuelled by a mixture of unsymmetrical dimethylhydrazine and nitrogen tetroxide. This was chosen because it burned spontaneously and required no ignition system. Two engines powered the first stage, whilst the second had only one engine, and steering was achieved by tilting the engines one way or the other.

To practice the rendezvous and docking procedures, an Agena D rocket was chosen because of its track record of excellent stabilisation and control whilst being used for the Ranger probes to the Moon, Mariner probes to Venus and Mars probes. Formerly the upper stage of the Thor and Atlas rockets, the Agena was fitted with a special adapter collar which contained a hydraulic system capable of absorbing the shock of another vehicle docking with it. Having done so, the Agena was able to lock it into place after the other vehicle had been drawn into position by motors. The locking mechanism was also integrated into the electrical system, which enabled the astronauts to issue commands to the Agena rocket and use the main engine to manoeuvre both craft.

To test this new vehicle the first two flights, Gemini I and II, were unmanned. Gemini I was used to test the compatibility of the Titan II rocket and the spacecraft itself. Both the vehicles were still mated when they re-entered the atmosphere. Gemini II was a similar experiment and climbed to an altitude of 160km before it too re-entered the atmosphere, again still mated to the Titan II rocket.

The development of the Gemini spacecraft and the successful unmanned tests of Gemini I and II culminated in the first manned flight, Gemini III, whose call sign was Molly Brown (taken from the Broadway play *The Unsinkable*), which was launched from a Titan II rocket at 0924 hours EST on 23 March 1965. It is said that Virgil Grissom's second choice for a name for the spacecraft was Titanic. This choice, as one can imagine, went down like the ship it was named after.

Things started well enough. The astronauts, John Young and Virgil 'Gus' Grissom, were strapped in the contoured couches 20 minutes before launch time, then, 35 minutes before launch, the countdown was halted. A first-stage oxidiser line had sprung a leak; a poorly seated nut had caused the problem. It was soon fixed and 35 minutes later the engines in the Titan II rocket burst into life. It was such a smooth lift-off that neither astronaut realised it was happening until they saw the mission clock running. But when the second stage engine ignited, the acceleration of the rocket brought them back to reality. The second stage engine shut down 5.5 minutes later and Gus Grissom kicked in the aft thrusters and pushed Gemini III into orbit.

During the flight some of the prepared 'space' food was eaten. John Young, however, produced a corned beef sandwich that fellow astronaut Wally Schirra had smuggled aboard for him. Passing it over to Gus Grissom, who had a couple of small bites, it was noted that the odd crumb floated about inside the capsule, something that raised a few concerns back in Mission Control. In fact, when some of the less supportive members of Congress got to hear of it, they voiced their dissatisfaction in no uncertain manner, and threatened all kinds of restrictions. Nothing more was said about the incident, but it did result in more stringent checks on what astronauts took aboard spacecraft on future flights.

As they approached their third orbit, Young fired detonators that separated the adaptor from the re-entry module and then, as they approached re-entry, the spacecraft slowed. First the drag 'chute was deployed, followed by the main 'chute, but as they made contact with the water the impact threw both astronauts forward. John Young's faceplate was badly scratched and Gus Grissom's was broken as they made contact with the windshield. The landing was made over 60 nautical miles from the pre-arranged landing point, so the crew had to await the arrival of a rescue helicopter.

With the two astronauts waiting in the stuffy capsule, Gus Grissom refused to crack the hatch remembering what had happened on his previous flight when he had been lucky to escape with his life as the spacecraft sank. As the temperature inside rose and the Atlantic swell increased, both men started to feel very uncomfortable. John Young, a navy man, managed to hold down his breakfast, but Gus Grissom, being an air force man, succumbed. Fortunately, after a 30-minute wait, a helicopter from the aircraft carrier USS *Intrepid* arrived and lifted the two men off their spacecraft and back to the ship.

Virgil Grissom and John Young had carried out a successful three-orbit flight, in which the spacecraft was fully evaluated by means of orbital manoeuvres using the Orbit Attitude and Manoeuvre System (OAMS) and a controlled re-entry flight.

Grissom and Young's flight was followed at 1015 hours EST on 3 June by the launch of Gemini IV with the call sign American Eagle. The crew, James A. McDivitt (Commander) and Edward H. White (Pilot), were woken at 0400 hours to prepare for their mission. After a physical examination and breakfast, the two astronauts left their Merritt Island quarters and headed toward the 'suiting up' area. There, Joseph Schmitt, the suit technician who helped them climb into their spacesuits, met them. Once inside, and with their helmets in place, they were put onto pure oxygen so that any nitrogen in their system could be removed. This would prevent aeroembolism, or the 'bends', as it is more commonly known.

Gemini IV patch.

Once inside their spacecraft, the two astronauts looked at the countdown clock, which registered T-100 minutes. The pre-flight checklist was methodically checked and they settled down for the launch. Then, at T-35 minutes, the signal occurred to which all the astronauts had become accustomed, the launch was put on hold. It appeared that when the erector was being lowered it suddenly stuck at a 12-degree angle from the booster. It was discovered on inspection that the wrong connector had been installed. In less than an hour the countdown was back on. At 1015 hours Gemini IV lifted off the pad at launch complex LC-19. This was to be a mission of firsts; the first mission to be broadcast to twelve European nations via the Early Bird satellite, the first to carry out an EVA (Extra Vehicular Activity) and the first to be controlled from the new purpose-built Mission Control at Houston, Texas.

As stage one separated, the explosive bolts that released the first stage shattered the relative quiet. This was followed shortly after by stage two separation, at which point the spacecraft went into orbit at a perigee of 163km and an apogee of 282km. As the spacecraft entered orbit, James McDivitt turned it around to look for the booster from which they had just separated. Activating the thruster's rockets that slowed the spacecraft down, he aimed it at the booster, but the booster for some unknown reason moved downwards. The crew tried again, but this time the booster seemed to be on a different track. They tried again several times and even during the night, guided by the flashing lights on the booster. This was to no avail, and when dawn came they saw the booster some 5km away. Jim McDivitt realised that he was using up more and more fuel in trying to rendezvous with the booster, so placed the ball squarely in Mission Control's lap. What was more important, the EVA (Extra Vehicular Activity) or the rendezvous? Back came the reply – the EVA.

Ed White started to prepare for the EVA, but whilst he was doing so McDivitt noticed that White looked strained and tired. Calling Mission Control, he asked that they do another orbit before Ed White carried out the spacewalk. A tired

Ed White carrying out America's first spacewalk using HHMU (Hand Held Manoeuvring Unit).

astronaut would not be in the peak of condition to take a walk in space, especially as it was a first, and no one knew what to expect. Mission Control agreed, and as the spacecraft passed over the Indian Ocean, Ed White, now rested and with his HHMU (Hand Held Manoeuvring Unit) gun in his hand, was ready. The tethered spacewalk was used to carry out a demonstration and evaluation of the EVA and the control of the HHMU. McDivitt de-pressurised the spacecraft and after some difficulty they managed to crack the hatch. Ed White floated out of the capsule and into the history books.

Outside the spacecraft, Ed White mounted a camera to record his EVA and then squeezed the trigger of the HHMU to propel himself away from the spacecraft. The sensation of floating freely in space was – in Ed White's words – sheer magic. He carried out turns, rolls and pitches, but then the compressed oxygen fuel bottle of the HHMU ran out. As the spacecraft orbited around the Earth with Ed White floating alongside, the astronauts carried out a radio broadcast to listeners all over the world, describing in great detail the beauty of the Earth as seen from space. Finally, after 15 minutes and 40 seconds, McDivitt had to ask White to come back inside the spacecraft. Reluctantly he agreed and the whole world heard him give a long sigh and say: 'It's the saddest moment of my life.'

Getting White back inside the spacecraft was not as easy as was first thought, but it was managed and McDivitt and White settled back in their seats for a well-earned rest. It was realised later that they had circled the Earth in an unpressurised spacecraft. Powering down the spacecraft, the two astronauts settled down to get some sleep. However, this was frequently interrupted by radio transmissions from Mission Control, so they decided to sleep in shifts. The remainder of the mission was taken up with observations, the taking of photographs (especially of weather phenomena) and experiments.

During the forty-eighth orbit of the Earth, Mission Control decided that the spacecraft's computer needed to be updated. McDivitt was told to switch the computer off as they flew over the United States, but he was unable to do so. Then, whilst trying out a number of other switch positions, the computer went dead. This caused a problem as the computer was to help them achieve a 'lifting bank' angle of attack for re-entry. Now they would have to resort back to a rolling Mercury-type of re-entry.

As they approached the sixty-second orbit they fired the manoeuvring thrusters for the required 2 minutes and 41 seconds, and then jettisoned the equipment adaptor. This was followed by the ignition of the retrorockets. As the

Ed White struggling to get back inside the spacecraft.

spacecraft reached 120,000m McDivitt commenced the rolling re-entry, slowing it down at 27,000m and deploying the drogue 'chute at 10,000m. At 1,500m the main parachute was deployed and the spacecraft splashed down into the ocean. Within minutes a helicopter was hovering overhead and two swimmers dropped into the water to fit a flotation collar around the spacecraft. The two astronauts were transferred to the helicopter and flown to the aircraft carrier USS *Wasp*.

Only one of the primary objectives of Gemini IV had not been accomplished and this was the computer controlled re-entry demonstration. This was because of an inadvertent alteration to the spacecraft's computer memory.

Later in the year, Jim McDivitt and Ed White went to the Paris Air Show where they were introduced to the Russian cosmonaut Yuri Gagarin, the first man in space.

On 17 February 1965, Ranger VIII was launched and sent back 7,137 pictures of the Sea of Tranquility, which was later to become one of the landing sites on the Apollo programme. The following month, Ranger IX, the last in the Ranger series, was launched and from the moment the television cameras were switched on, the whole programme was live on television, bannering a headline across the world's television sets 'LIVE FROM THE MOON'. As exciting as these missions were, they still left a large gap in the programme of putting a man on the Moon. All the photographs taken of the Moon's surface did not give NASA the information they needed before putting a spacecraft onto the surface.

It was left to the Russians to make the next step, when in January 1966, after a series of failed attempts, Luna 9 soft-landed on the surface of the Moon. Eight hours later its television camera was switched on and started to send back pictures. The quality of the pictures left a lot to be desired, adding fuel to the speculation that Russian space technology was struggling to keep pace with progress, but it gave the Americans some of the information they wanted – man could walk on the lunar surface. In March, Luna 10 was put into orbit around the Moon, giving the Moon its first artificial satellite. From their remarkable start, the Russians seemed to have lost their impetus, as the technology and finance required to put a man on the Moon seemed beyond them. They even shelved plans to build a spaceplane called the Raketoplan, which had been designed by the Russian designer and manufacturer Chelomei as a space interceptor.

Another breakthrough in space travel was made on 22 February 1966 when Voskhod 3 was launched carrying two dogs Verterok (Little wind) and Ugolek (Little piece of coal). The spacecraft with its two occupants went into orbit around the Earth for twenty-two days before landing safely back home. Their space record was not surpassed until the flight of Skylab-2 in June 1974.

Two months later on 30 May, the Americans launched Surveyor I from Cape Canaveral and landed the spacecraft in the Ocean of Storms region, sending back 11,240 pictures of the Moon's surface. Three months later, the first Lunar Orbiter went into orbit around the Moon, orbiting at 160 miles (257km) above the surface. This was the first of the Lunar Orbiter series whose job it was to map the Moon's surface in preparation for a manned Moon landing. All the Orbiters had a miniature photographic laboratory built into them, and each was capable of processing 200ft of film. Four more Orbiters were sent to map the Moon within the following year. Meanwhile the Surveyor project continued. Surveyor II crashed on landing, but Surveyor III landed safely in the same region of Surveyor I (the Ocean of Storms), sending back 6,315 pictures and information on the structure of the lunar surface. Surveyor IV worked perfectly until just after landing in the Sinus Medii region, when, for no apparent reason, it just shut down. Three more Surveyor spacecraft landed on the Moon, all sending back pictures, 69,000 in total, and other information.

The next manned spacecraft to blast off the pad at Launch Complex LC19 was Gemini V at 0900 hours EST on 21 August 1965. The crew, Charles 'Pete' Conrad (Commander) and Gordon Cooper (Pilot), were about to go on the longest manned spaceflight to date and to complement this, Gordon Cooper had a crew patch made that depicted a covered wagon with the motto '8 Days or Bust'. NASA's administrator James E. Webb, not noted for his sense of humour, was not amused. Conscious of the difficulties that NASA experienced when looking towards Congress for funding, he turned it down, fearing that if the mission were to be aborted early then many would say that it had 'busted', lending a degree of flippancy to the project, which was the last thing he wanted levelled at NASA. He then accepted the patch on the conditions that the '8 Days or Bust' was removed. The Gemini V mission patch was the first ever to be made by a crew and every crew from that day on designed their own to symbolise their mission.

After the Gemini V spacecraft had entered into orbit around the Earth, at an apogee of 193 miles (349km), Conrad and Cooper settled down to carry out some of the medical experiments. These consisted of Cardiovascular Conditioning and the Human Otolith Function, designed to see if horizontal perception deteriorated during spaceflight. Boredom was the biggest enemy to the crew. Because of the restricted movement within the capsule, experiments were very limited, but photography was one of the things that could be carried out easily.

A number of problems started to come to light during orbit. One of them concerned the supply of oxygen and hydrogen for the fuel cell. Conrad had noticed that the pressure in the fuel cell had dropped too low and Mission Control advised him to switch on the oxygen heater in an effort to raise the pressure, but the problem was to get worse. Two hours 13 minutes after launch, Cooper ejected the rendezvous pod then, turning the spacecraft to the rear, he switched on the radar picking up the signal from the pod immediately.

The pressure drop in the fuel cell was now becoming a cause for concern. It had dropped to below 200lb per square inch and neither astronaut had seen a fuel cell working at such a low pressure before. The decision had to be made either to power down the electrical systems or run the risk of the fuel cell stopping entirely. Rendezvousing with the pod was now out of the question, but if the trouble could be sorted then maybe they could carry out some simulated trials. Back in Mission Control, Flight Director Kraft knew that there was enough power in the batteries for the spacecraft to carry out a re-entry even if the fuel cell was to fail completely, but it was the re-entry zone that was causing him concern. So the electrical systems were powered down whilst the problem was sorted out. It was decided to continue as planned and Cooper turned the electrical systems back on, but during the power down, whilst the spacecraft drifted in space, the capsule had become very cold. This caused the astronauts to have great difficulty in sleeping. They tried sleeping alternately, but as one slept the other had to work, and consequently the one sleeping kept being woken by radio transmissions from Mission Control. So they decided that they would both work, eat and sleep at the same time, and although they still had difficulty, it was considerably easier.

Because they were unable to carry out the rendezvous with the target vehicle, the Gemini V crew carried out simulated manoeuvres. All these were successful and Cooper was able to place his spacecraft in the right position every time. Then on the fifth day another problem struck the spacecraft. The Orbital and Manoeuvring System (OAMS) became sluggish and one of the thrusters became inoperative. Six hours later a second thruster shut down and the spacecraft started to tumble through space. The electrical system was shut down again and all other experiments that required fuel were cancelled. However, of the seventeen experiments that did not require fuel that had been assigned to the crew, only one was eventually cancelled. Conrad and Cooper continued to take photographs of both the stars and the Earth.

On the morning of 29 August, 190 hours and 27 minutes after leaving Earth, Gemini V fired her retrorockets and began to re-enter the Earth's atmosphere over Hawaii. At 6,000ft Cooper deployed the drogue parachute, seconds later the main parachute deployed and minutes later the spacecraft splashed down into the sea. Despite the problems, Gemini V had been quite successful and had proved that the human body was quite capable of adapting to the relatively long-term weightless conditions in space.

The next planned mission was to have been the first flight to dock with an Agena target/propulsion stage, but the stage blew up on its way into orbit. So it was decided that Gemini VI would carry out a rendezvous with Gemini VII. This mission would require a one-week turnaround. The two spacecraft would have to be launched just over a week apart, something the Americans had never tried before. They had always allowed a minimum of three to four weeks between launches. Gemini VII, with astronauts Frank Borman and Jim Lovell aboard, was launched from Cape Canaveral on 4 December 1965 ahead of Gemini VI. Initially the idea was for the two spacecraft to rendezvous in orbit and the two pilots to carry out a spacewalk over to the other's spacecraft and resume each other's roles. The idea behind the proposal was to set in motion the beginnings of an exercise that could be brought into play in the event of an accident happening in space and a rescue mission having to be carried out. Borman and Lovell talked long and hard with the other astronauts who had spent long periods of time in space, learning about the pitfalls such as sleep deprivation, boredom and fatigue. They decided that they would share the same work and sleep periods and made that clear to the Mission Controllers. All experiments and tasks would be carried out at appropriate times, not just one after another.

Six minutes after lift-off, Gemini VII cut loose from its booster, followed 2 minutes later by separation from the second stage. The two astronauts kept a wary eye on the second stage that was jumping all over the sky as it spewed out its remaining fuel. Borman fired the spacecraft's thrusters and moved into orbit. Interestingly enough the booster was still in sight with its recognition lights still flashing off a myriad of particles that surrounded it. Then, about 6km away, both Borman and Lovell reported what they could only describe as three unidentified objects in orbit. Neither party ever mentioned the subject again. They then reported that a fuel cell warning light had come on, but Mission Control, after some debate, decided that it was of no consequence and they would just keep an eye on it.

The following day the two astronauts settled down to carry out a variety of experiments, which were all successful. The cramped capsule was now becoming rather stuffy, so Lovell took off his pressure suit, which eased the situation somewhat. They had both decided to remove their pressure suits, but Mission Control vetoed it, saying that one of the astronauts should have a pressure suit on at all times. It was becoming a problem, even though the suits were relatively lightweight, because of the high temperature and the confined space the two astronauts had to work in. They contacted Mission Control requesting that both of them be allowed to remove their suits, but again they were told no. Borman and Lovell changed roles; this time it was Borman who sat in suitless comfort, whilst Lovell sweltered. Relief came when they were told that the Director of the Manned Spaceflight Center had given permission for both of them to take off their pressure suits.

More experiments and general information were relayed to the astronauts over the few days, and then on 15 December 1965 Gemini VI with astronauts Wally Schirra and Tom Stafford on board, blasted off the pad at the Kennedy Space Center. After the two stages had separated without incident, Gemini VI entered orbit. Gemini VII had been passing over the Cape area at the time, but had seen nothing because of cloud. Then as they passed over the Tananarive, Malagasy Republic, they saw the contrails of Gemini VI approaching orbit. It was time to get their pressure suits on, as the task of manoeuvring to a rendezvous was about to start.

On Gemini VI, Tom Stafford had been told that he was 1,992km behind Gemini VII; 94 minutes later and in a slightly lower orbit, Stafford kicked in the thrusters to speed up his spacecraft – they were closing fast. It was to be 5 hours and 16 minutes before they were able to catch up with Gemini VII and then they were close enough to carry out the delicate task of manoeuvring the two spacecraft by the gentle pressure of the reaction thrusters. Closer and closer the two spacecraft got until they were a mere 2m (approx 6ft) apart. The skill that got the two spacecraft together was about to be tested further by 'stationkeeping' the two spacecraft. This meant that the two spacecraft had to maintain the same speed and the same distance as they orbited the Earth. After several more burns the two spacecraft were only 40m apart. The burns had only used 51kg of fuel on Gemini VI, leaving plenty for some fly arounds. During the next 270 minutes the crews moved as close as 30cm, talking over the radio. At one stage the spacecraft were stationkeeping so well that neither crew had to make any burns for 20 minutes.

The two spacecraft parted after twenty orbits, and Borman and Lovell prepared to carry out pre-entry checks. These checks took over 2 hours. On 18 December

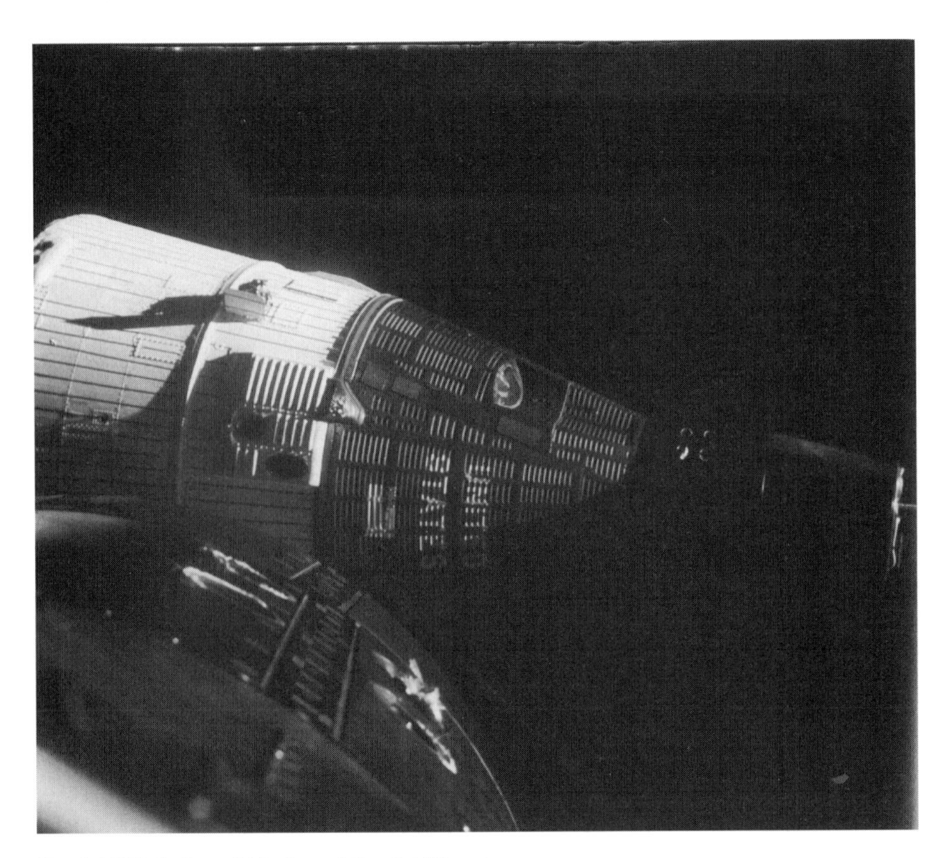

Gemini VII closing within feet of Gemini VI.

Borman fired the retrorockets as they started re-entry, then the computer took over and held the spacecraft steady until the drogue 'chute went out. This was followed by the main 'chute, which steadied the spacecraft until it impacted on the water. Half an hour later the two astronauts were on the deck of the aircraft carrier USS *Wasp*. There they were joined by Wally Schirra and Tom Stafford, whose spacecraft Gemini VI had landed two days earlier on 16 December.

The Gemini programme continued with the launch of Gemini VIII, with astronauts Neil Armstrong and David Scott (both destined to later walk on the Moon) on board. This was to be the next attempt at docking in space with the Agena target vehicle. The Atlas-Agena combination lifted off the pad and into orbit, much to the relief of those in Mission Control. The powerful Titan II engines burst into life and Gemini VIII lifted off the pad at Launch Complex LC19 with a roar into the skies, a perfect lift-off. By the time Gemini VIII arrived in orbit it was about 1,963km behind the Agena target vehicle. As the spacecraft reached their second orbit all the initial checks had been done, so Armstrong fired the thrusters to increase the spacecraft's speed and cut down the distance between them and the Agena target vehicle. At 332km their radar suddenly locked on to the Agena. As Gemini VIII closed with the target vehicle they suddenly made visual contact. Armstrong and Scott started to make final preparation. Inch by inch they edged closer until they were less than 2ft apart, at which point the two craft maintained station. Armstrong kept the two craft apart for 36 minutes whilst they checked the Agena over for any damage that might prevent the docking.

Easing his spacecraft forward, Armstrong docked Gemini VIII, the first time two spacecraft had docked whilst in space at an incredible speed of 17,500mph. The euphoria was short-lived as suddenly the situation began to deteriorate. The two spacecraft started to roll, slowly at first, but gaining momentum all the time until they were spinning at one revolution per second. After trying to stabilise the two spacecraft, they decided to separate. Gemini VIII pulled away from Agena, but the rolling continued and nothing seemed to prevent it. Then Armstrong tried the hand controllers and fired the rear thruster on his spacecraft, which solved the immediate problem. Later, by a process of elimination, they discovered that an electrical wiring fault had caused the No 8 yaw thruster engine to malfunction. An override master switch was installed on all future Gemini spacecraft to prevent this from happening again.

Despite the thruster problem, the docking exercise itself was a complete success and the remainder of the mission was taken up by experiments. But the docking manoeuvres had taken their toll on the fuel and Mission Control ordered the spacecraft to abort the planned spacewalk by Scott and start the re-entry procedure. The crew went through the pre-re-entry checklist and then started re-entry. Ten hours and 41 minutes after they had launched from Cape Canaveral, the spacecraft splashed down into the Pacific Ocean just 2 miles from the planned landing point. It was to be over an hour before the first of the rescue planes sighted them in the rough seas. Three para-rescue team members jumped in and attached a flotation collar around the spacecraft. Then the destroyer USS *Leonard F. Mason* drew alongside and hauled the astronauts and their spacecraft aboard.

o started off on a tragic note. The prime crew, astronauts Elliot M. See and Charles A. Bassett, were killed when their T-38 jet trainer crashed in rain and fog just short of landing at St Louis Municipal Airport. The aircraft was left of centre on its approach to the main runway and hit the roof of one of McDonnell's buildings. Minutes later Thomas Stafford and Eugene Cernan, who were back-up crew for Gemini IX, landed safely in their T-38 trainer. Both crews had been en route to McDonnell for two weeks' training in a simulator. Stafford and Cernan were immediately upgraded to prime crew and an investigation was set up to examine the cause of the accident. It was concluded that it was pilot error coupled with very poor visibility.

On 3 June 1966, Gemini IX blasted off the pad at Cape Canaveral with astronauts Eugene Cernan and Tom Stafford aboard. The mission had started on 17 May with the launch of the Agena Target Vehicle from the Kennedy Space Center, but shortly after launch the rocket turned on its side, headed down and crashed into the Atlantic. The launch was postponed until 1 June when another Atlas Launch Vehicle, which had been prepared for the mission, was launched. The launch was perfect but then a problem was discovered. It was suspected that a launch shroud that covered the docking port had failed to jettison.

Gemini IX entered orbit and commenced to catch up with the Agena Target Vehicle. The crew, in the meantime, was carrying out pre-rendezvous procedures and a variety of small experiments. At a range of 240km, their radar picked up the Agena, and then at 93km they had a visual sight of their target. It was not until they were only 30m apart that Tom Stafford could see that the shroud was still in place. 'It looks like an angry alligator' was Stafford's initial reaction, but he quickly realised that there was no way that they were going to be able to carry out a docking manoeuvre. Stafford and Cernan even considered trying to nudge the shroud off the Agena using the spacecraft, but dismissed it as being too risky.

To try and get a closer look at the problem, Tom Stafford moved his spacecraft to between 9–12m away and rolled it in parallel with the movement of the Augmented Target Docking Adaptor (ATDA). It was a manoeuvre that was to be used in later flights to examine unidentified satellites. Stafford could see that although the explosive bolts had fired, two lanyards of high tensile steel still partially held the shroud in place.

At a post-mortem later it was discovered that the fault lay ostensibly with the manufacturers and assemblers of the shroud and the ATDA. In short, too many different people were involved and communication between them was woefully lacking.

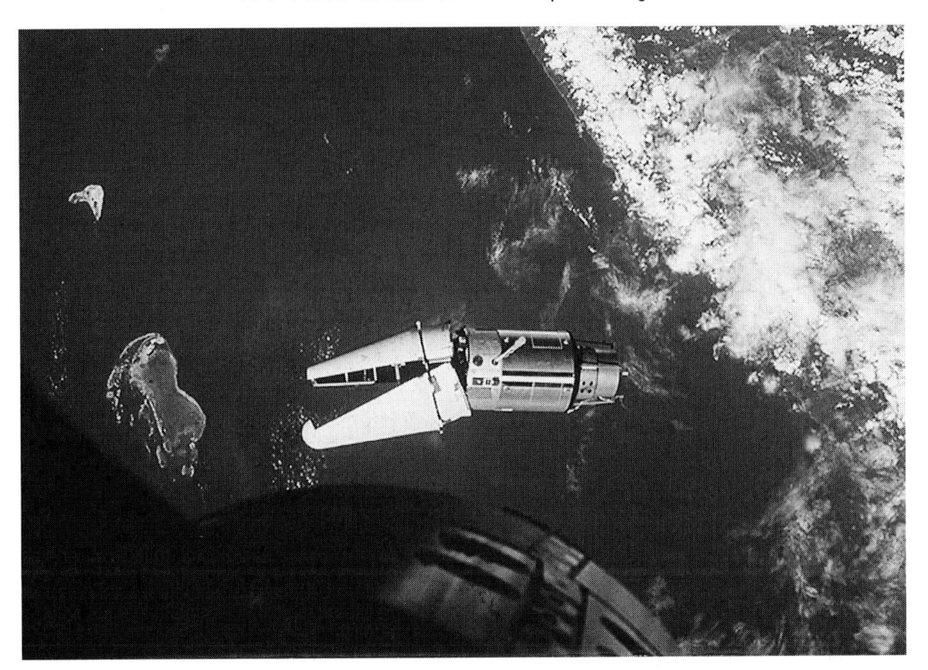

Shot of the Agena Target Vehicle with its shroud still partially intact which prevented the docking with Gemini IX to take place.

The crew of Gemini IX was now reduced to carrying out stationkeeping and rendezvous exercises, which in themselves were extremely useful. But the fuel level was beginning to worry Stafford so it was decided to call a halt to the mission, and on 5 June Eugene Cernan carried out his spacewalk. Ed White had commented on his spacewalk that there was a need for hand holds around the spacecraft. But even with these in place Cernan found that the very act of moving around the spacecraft was exhausting. He attached himself to the manoeuvring unit and attempted to carry out his tasks. Then his visor started to fog up and this restricted his view, making use of the manoeuvring unit a waste of time and energy. After completing over three-quarters of his tasks, Cernan decided that it was now becoming almost impossible to see because of condensation. Releasing himself from the manoeuvring unit, he groped his way back to the cockpit where Stafford helped him into his seat. Stafford could not see Cernan's face through his faceplate, even though their helmets were touching.

The rest of the day consisted of carrying out experiments, one of them for the Department of Defence (DoD). On 6 June, and after the forty-fifth orbit of the Earth, Gemini IX splashed down into the Atlantic Ocean. The spacecraft and its two occupants were hoisted aboard the aircraft carrier USS *Wasp* to a hero's welcome.

Gemini IX being stabilised by frogmen with an aircraft carrier closing with them to recover the spacecraft and the astronauts.

A number of things had been learned from this flight and amongst them was the problem of the fogging up of the faceplate whilst on EVA. Had Eugene Cernan not been tethered to the spacecraft, who knows what might have happened.

One of the most successful Gemini missions was that of Gemini X. The crew, Michael Collins and John Young, were launched from Cape Canaveral on 18 July 1966. One hour and 40 minutes earlier the Agena Target Vehicle had been blasted off the pad on top of an Atlas rocket, attaining a near-circular orbit, 162-mile apogee and 157-mile perigee. Gemini X left on time and went into orbit trailing behind the Agena by some 1,000 miles and at an initial apogee of 145 miles. John Young quickly activated the thrusters and caught up with the Agena Target Vehicle on the fourth orbit, some 5 hours and 23 minutes after launch. Lining up his spacecraft, he had to make two mid-course corrections before finally docking. More propellant had been used during the docking manoeuvre than had been estimated, so a number of other manoeuvres were cancelled. It was decided to keep the spacecraft docked with the Agena Target Vehicle in an effort to save fuel, and alter the mission plan to suit. Several alternative experiments were carried out including EVAs, and during the 39 hours in which they were docked six manoeuvres were carried out to put the spacecraft and its ATV into a position to rendezvous with the ATV of Gemini VIII.

The Gemini X spacecraft undocked after 44 hours and 40 minutes and Collins carried out a 39-minute EVA to recover a micrometeorite package that had been attached to the ATV of Gemini VIII. As Collins approached the ATV he stretched out to grasp one of the handholds but lost his grip on its smooth surface and started to drift away from it. Using his HHMU Collins propelled himself back and retrieved the package. Getting back into the spacecraft proved extremely difficult because the umbilical cord had wrapped itself around him and he had to extricate himself with the help of John Young.

The spacecraft stayed on station-keeping duties for a further 3 hours while the crew prepared the spacecraft for re-entry. After their forty-third orbit of the Earth and 70 hours and 10 minutes in space, Collins and Young carried out a retrofire manoeuvre and splashed down within sight of the aircraft carrier USS *Guadalcanal*.

As per the previous two launches, the Gemini XI mission began with the launch on 12 September 1966 of the Atlas-Agena Target Vehicle. One hour and 40 minutes later, Gemini XI, with astronauts Charles 'Pete' Conrad Jr and Richard F. Gordon on board, blasted off from Cape Canaveral. Their prime objective was to catch up, rendezvous and dock with the ATV within the first orbit. The two spacecraft rendezvoused after 1 hour and 25 minutes and, after five manoeuvres, docked 9 minutes later.

Earlier in the Gemini programme, Conrad had been aware of a proposal of a Gemini flight to orbit the Moon. The NASA hierarchy had dismissed the idea because plans were well underway for the Apollo missions, which were being designed specifically for a flight to the Moon. Conrad reluctantly accepted this, but argued that they should take a Gemini flight out into space as far as they could so that they could take colour photographs and film of the weather systems that surround the Earth. The Weather Bureau had long since complained that their high-flying weather satellites were sending back poor quality black-and-white pictures and were seriously considering the use of colour. If Gemini XI was allowed to go out deeper into space and take colour pictures, then there might be an argument for the Weather Bureau to use a colour system.

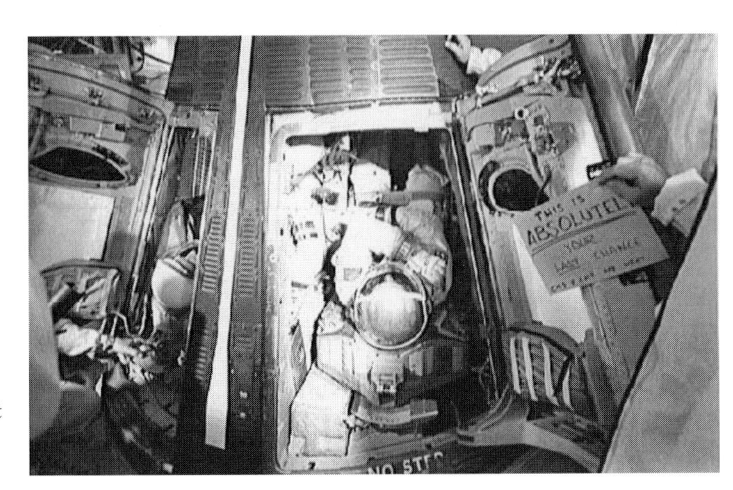

Gemini XI crew, Pete Conrad and Richard Gordon in their spacecraft just prior to launch.

The crew of Gemini XI got down to carrying out the experiments and trials assigned to their mission and amongst these was an objective to tether the Gemini spacecraft to the ATV. The idea being that in future spaceflights, when spacecraft were sent to repair satellites they could tie up alongside. There were problems with the tethering because it had to be decided how long the tether should be to prevent over-tensioning when the spacecraft rotated. They also carried out docking and undocking procedures in daylight and in darkness, something that NASA had wanted to do for some time.

Gemini XI docked with the ATV and then 'lit' Agena's Primary Propulsion System (PPS) to put them into a higher orbit of 741 miles. Settling into their new orbit, the two astronauts decided to have a meal then try to get some rest before embarking on the next objective, Richard Gordon's EVA.

The two men got suited up ready for the vacuum environment and waited for the scheduled time. As the sun came over the horizon, Gordon began putting a sun visor over his faceplate. Conrad helped with the left side but Gordon struggled to fasten the right-hand side, and as he snapped it into place, it cracked. Nevertheless they opened the hatch and Gordon floated out into space, his tethering line snaking out behind him. He drifted over to the ATV with the intention of attaching a tethering line during his planned 1-hour EVA. After struggling for nearly 10 minutes, Gordon secured the line. It suddenly became obvious to Conrad that all the training in the zero-G aircraft was relatively easy compared

Gemini XI closing with the Agena Target Vehicle.

to what Gordon had to do. In the aircraft all they wore were flight suits, in space they had to wear cumbersome EVA suits. Gordon was running into difficulty physically; sweat was pouring down his face and into his eyes, half blinding him. After 30 minutes of watching his struggle, Conrad called a halt to the EVA and ordered Gordon back into the spacecraft. After closing the hatch and pressurising the spacecraft, Conrad helped Gordon to undress, seeing how exhausted his pilot was. There was to be one more stand-up EVA, but that would not cause the same kind of problem that Gordon had just experienced.

After the twenty-eighth orbit of the Earth, Conrad brought the spacecraft back down to an apogee of 164 miles and started preparations to undock. After some problems releasing the tethering line, they undocked from the ATV and moved away from the now defunct vehicle. They now placed their spacecraft in an orbit above the ATV and carried out an exercise in long-range station keeping. Whilst carrying out this exercise, they began testing a night image intensifier. This piece of equipment was designed to see if their night vision would be improved by allowing them to scan objects on the ground, relaying back the pictures to a monitor in the spacecraft. Both agreed that this was probably one of the more enjoyable and successful of tests that they had to undertake.

They then carried out a second rendezvous manoeuvre with the ATV, which was successful. The crews were now in a relaxed frame of mind and were keen to carry out more exercises, but their mission was over and the only thing left was the automatic re-entry. This was a first, because up to now all re-entry manoeuvres had been manually controlled, but this one was to be controlled by computers. After the forty-fourth orbit of the Earth, the retrorockets fired and the spacecraft entered the atmosphere right on schedule, splashing down in the Atlantic Ocean close to the aircraft carrier USS *Guam*.

Lunar Orbiter II was launched on 6 November 1966, to orbit the Moon to look for possible landing sites for the forthcoming Apollo missions. The writing was on the wall for the Gemini project.

When astronauts Edwin 'Buzz' Aldrin and James A. Lovell walked up the ramp to their waiting spacecraft, the signs on their backs said 'THE' and 'END' and the end it was for the Gemini programme. Gemini XII was to be the last of the Gemini missions and it was hoped that all the mistakes and problems would be resolved on this the last flight. The end of Gemini could hardly have been clearer; as soon as the Titan II rocket lifted of the pad, the wrecking crews were there demolishing what would have been part of the history of manned flight.

The launch of Gemini XII was not without its own problems. Two instruments in the launch vehicle had to be replaced: a secondary autopilot and a Stage I rate gyroscope. Shortly after they had been installed, they too developed faults and had to be replaced. Not the auspicious start Mission Control had anticipated. But on 11 November 1966, Gemini XII lifted off the pad to rendezvous in orbit with the Atlas-Agena Target Vehicle, which had been launched a couple of hours earlier.

Jim Lovell and Buzz Aldrin, with the words 'The' and 'End' on their respective backs, head towards boarding their Gemini XII spacecraft.

Once in orbit, Gemini XII quickly locked on to the ATV, which was 236 miles ahead. After an adjustment to send the spacecraft into the same orbit as the ATV, the radar went out. Fortunately Aldrin, ever the meticulous one, had worked out chart procedures in the event of the radar going faulty and now he was about to put his calculations to the test. Using a sextant, Aldrin calculated the range and range rates and fed the numbers to Jim Lovell. After 3 hours and 45 minutes of flight, the spacecraft rendezvoused with the ATV and 30 minutes later announced jubilantly 'We are docked'; a testament to the skill of Aldrin and Lovell.

Aldrin had listened carefully to the reports of the other astronauts that had carried out EVAs and the problems they had encountered. Armed with this knowledge, he opened the hatch of Gemini XII just 20 minutes before sunset. Even he wasn't ready for the awe-inspiring sight that greeted him. As the spacecraft drifted into night, he let his eyes become accustomed to the dark and then looked in wonderment at the uninterrupted view of the starts and planets that shone untwinkling above him. Setting up an ultraviolet astronomical camera, he set about photographing the star fields around him. He later recovered a micrometeorite collection package from the spacecraft and then returned to the capsule after carrying out a 2-hour and 20-minute EVA.

The following morning Aldrin got ready to carry out another EVA, only this time it was to be on the end of an umbilical cord and a number of tasks had to be completed. Leaving the spacecraft, he moved slowly hand over hand along the handrail to the nose of the Agena docking adaptor. Aldrin knew that any attempt at rapid movement in a weightless environment would cause exertion, which he knew from the other astronauts, was debilitating. Tethering the two spacecraft together, he prepared them for the gravity-gradient experiment. Moving along the spacecraft, Aldrin practised using a torque wrench and fitting electrical connectors together. The EVA lasted 2 hours and on the way back in, Aldrin even had time to wipe the windscreens over with a dry cloth. This was accompanied by comments from Lovell: 'Have you checked the tyres and oil?'

Another EVA was carried out the following day, but this time it was only a stand-in-the-seat exercise and just before re-entering they discarded all the unwanted equipment and garbage out into space. All these items and rubbish would continue to go around the Earth in a decaying orbit, burning up as they entered the atmosphere.

There had been a number of problems on this final mission, but it had accomplished nearly all the objectives that had been set and the information gathered was priceless. As the spacecraft entered its fifty-ninth orbit of the Earth, the retro-rockets were fired automatically and Gemini XII began its descent into the Earth's atmosphere. As it did so, a pouch that contained a variety of small equipment and charts broke loose and landed squarely in Jim Lovell's lap. He resisted the urge to grab it because the two astronauts had activated the ejector seat D-rings and Lovell was terrified that if he clutched at the pouch he might inadvertently grab the D-ring and yank it upwards. The thought of ejecting into the atmosphere as they were plummeting through it was something he did not relish. The spacecraft splashed down in the sea just 3 miles from the recovery aircraft carrier USS *Wasp*. The Gemini programme was over and another chapter in space exploration was closed, but the next was about to open.

The Americans, now having finished the Gemini project, put all their thoughts and resources into putting a man on the Moon. But first they had to get more information about what they were going to land on. It was with this in mind that the Ranger flights were developed. The first six Ranger flights were a disaster, but the seventh, Ranger VII, blasted off, and 63 hours later as the spacecraft approached the Moon six television cameras on board were switched on. Sending back pictures at a rate of one every 2.5 seconds, the spacecraft hurtled toward the Moon's surface before crashing into one of the many craters that pockmarked it. The detailed pictures were superb and gave NASA an insight into what lay before them.

Disaster struck the Russian space programme when, on 23 April 1967, Soyuz (Union) I was launched with Cosmonaut Vladimir Mikhailovich Komarov on

Vladimir Mikhailovich Komarov.

board. Just after orbital insertion things started to go wrong. First one of the solar panels failed to deploy, wrapping itself around the Service Module, reducing the spacecraft's power by 50 per cent. This then prevented the spacecraft from manoeuvring. Two hours before the decision was made to bring the spacecraft back to Earth, Ground Control knew they had serious problems. The tragedy was that Komarov knew the situation to be dire and was well aware that his chances of survival were slim. The gravity of the situation was made more apparent when Premier Kosygin called Komarov personally on a videophone link and was heard to be crying. Komarov's wife also came on the link-up and Komarov dictated to her how she was to take care of the children and his affairs. Re-entry was made safely, and at the correct altitude the drogue 'chute deployed. Then, because of a faulty pressure sensor, the main 'chute did not deploy, so Komarov released his reserve 'chute manually, but the parachute straps got twisted with the drogue

Said to be the remains of Vladimir Komarov in his coffin, being examined by fellow officers.

'chute. US monitoring stations in Turkey picked up the last few moments of Komarov's life as he ranted and raged at the men who built the faulty spacecraft. The spacecraft crashed into the ground with the force of a 3-ton meteorite, near the town of Orenburg, killing Komarov instantly.

Komarov's ashes were buried in the Kremlin wall with great pomp and ceremony. This put the Russian space programme on hold until the cause of the accident was found and all their other spacecraft could be modified to prevent it from happening again.

There had been reservations about the flight some months before, with over 200 known defects in the new Soyuz spacecraft. Komarov knew of these and expressed a reluctance to go on the flight, but realised that Yuri Gagarin, being the back-up pilot, would have to take his place. Just before the flight, Gagarin and a number of the other cosmonauts prepared a document detailing the defects and requesting that the mission be aborted. The Russian hierarchy, however, wanted desperately to have something special to celebrate the fiftieth anniversary of the Russian Revolution, so the document never left the KGB headquarters. The rest, as they say, is history.

It was rumoured that the designer of Soyuz 1's parachutes, Pavel Tkachev, was imprisoned for two years after Komarov's death. Because it was almost impossible to pinpoint which individuals were responsible for the 203 defects known to have been in the spacecraft, the designer of the parachutes was the obvious and easiest one to hold responsible.

CHAPTER FOUR

THE RACE TO THE MOON

On 8 December 1966, the United States and Russia agreed a treaty that limited the use of military activities in space. Then on 27 January 1967 United Nations members signed the space law articles, 'Treaty on Principles Covering the Activities of the States in the Exploration and Use of Outer Space, Including the Moon and other Celestial Bodies'. This auspicious start was marred by a fatal accident at the Kennedy Space Center.

The Apollo programme had started on 26 February 1966 with the launch of Apollo-Saturn 201 (AS-201). This was an unmanned flight designed to test the separation of the various stages that made the Saturn V rocket. It was also to check the launch escape systems, the guidance, propulsion and electrical subsystems of the launch vehicle, the heatshield of the spacecraft Command Module (CM) and recovery of the CM.

On 5 July the second of the unmanned Apollo rockets, Apollo-Saturn 203 (AS-203) was launched. This was to check out:

a. Venting and chill-down systems,
b. Fluid dynamics and heat transfer to propellant tanks,
c. Attitude and thermal control systems,
d. Launch vehicle guidance,
e. Checkout in orbit.

The following month the Apollo-Saturn 202 (AS-202) was launched, to examine and evaluate the following:

a. Command Module (CM) heatshield at a high heating load,
b. Structural integrity and compatibility of the launch vehicle and the spacecraft,
c. The flight loads,
d. Stage separation,

e. Subsystems operations,
f. Emergency detection system operations.

Then, on 27 January 1967, tragedy struck. Three astronauts, Virgil 'Gus' Grissom, Edward White and Roger Chaffee, were in the Command and Service Module (CSM) aboard Apollo I, which was to be the first manned spacecraft of the Apollo programme. The CSM was on top of a Saturn V launch rocket, which was on launch complex 37 at the Kennedy Space Center and was going through a simulated countdown under launch conditions. Suddenly over the intercom came a shout:

'Fire!'
'We've got a fire in the cockpit!'
'We've got a bad fire … Let's get out. We're burning up …'

Seconds later this was followed by a piercing scream. It was determined later that a spark must have ignited the pure oxygen atmosphere in the capsule and within a matter of 12 seconds the three astronauts were dead. Although the men were wearing spacesuits, the inside of the capsule had been turned into a raging inferno, estimated to be in excess of 1,000°C in less than 10 seconds.

The last voice transmission before the fire was recorded at 23:30:14 GMT. The next transmission occurred at 23:31:04 when the fire was reported. Emergency procedures both inside the capsule and outside were activated, but such was the rapidity of the fire that within seconds the capsule had ruptured and flames, gases and thick black smoke were pouring out of the capsule and filling the White Room. By the time rescuers had reached the area it was impossible to see through the smoke and all movements had to be made without the aid of masks and by touch.

The Apollo insignia.

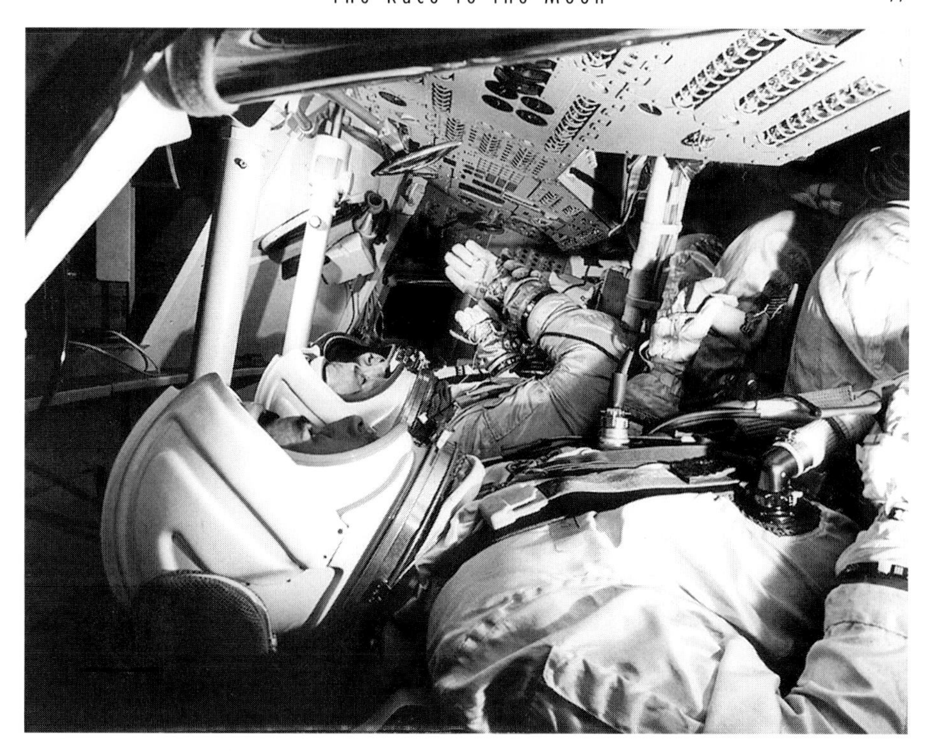

The crew of Apollo I, Grissom, Chaffee and White, in the simulator.

Because the Command Module was connected to the swing arm which was atop the gantry, the Command Module (CM) had three hatches installed. The outermost hatch, known as the BPC (Boost Protective Cover), protected the CM during launch and was jettisoned just prior to orbital insertion. The second and middle hatch, which became the outer hatch after launch, was covered with an ablative material to protect the astronauts during re-entry. The third and inner hatch created the pressure seal. The outer hatch was not fully closed during the tests because of large wire looms that were connected to the instruments for monitoring purposes. But in order to open the hatches from the outside, a specially designed hand tool was used to create a handhold, and this tool was situated in the White Room.

With the smoke seriously limiting visibility, it took minutes to locate the tool and even longer to get the first of the three hatches open, mainly because of the breathing difficulties the rescuers were having. When the inner hatch was finally opened thick heavy smoke poured out accompanied by intense heat. In all, the time taken to open all three hatches was 6 minutes, but it was a further 7.5 hours before the smoke and heat had subsided enough to enable the removal of the three astronauts' bodies. This sad task took over 90 minutes such were the conditions inside the cramped capsule.

Remains of the interior of Apollo I capsule in which the three astronauts died.

Lessons were quickly learned, but it was never established beyond doubt what actually caused the fire. Any evidence that could have determined the cause was destroyed by the ferocity of the fire. But by a process of elimination, one of the main causes put forward was a wiring fault beneath the left-hand couch (Gus Grissom's), and the fact that the atmosphere in the capsule was pure oxygen.

The main concern was to ensure that such an accident did not happen again, and several factors were identified that contributed to the tragedy. The ingress/egress hatch opened inwards and this was redesigned to open outwards. A number of other alterations were made to the design and construction of the Command Module before the subsequent crews were destined to step inside it. The next designated crew to fly consisted of Walter Schirra, Don Eisele and Walter Cunningham. Schirra had flown on both the Mercury and Gemini flights and, consequently, was a very experienced astronaut. To ensure that all the new safety features were being incorporated into the new Command Module, he virtually slept alongside the module at North American Rockwell's facility and checked the workmanship constantly, much to the chagrin of some of the construction crews. His only answer to the critics was, 'You only have to build it. My crew and I have to live in it and our lives depend on it'.

NASA also announced that the atmosphere inside the spacecraft would no longer be 100 per cent oxygen whilst on the launch pad, but 60 per cent oxygen and 40 per cent nitrogen. The 100 per cent oxygen atmosphere would only be retained whilst the capsule was in space.

Just less than a year later, on 9 November 1967, Apollo IV (Apollo-Saturn 501) was launched from Launch Complex 39. This was the complete unmanned Saturn V rocket with spacecraft attached consisting of a Lunar Module Test Article (LTA) and Command and Service Module (CSM). It was to carry out a full evaluation before the first manned flight took place. Not only was the launch vehicle to be fully evaluated, so was the spacecraft itself. This was to be one of the most intensive evaluations of the programme.

The first stage cut-off occurred 2 minutes and 30 seconds after launch at a height of 63km. The second stage ignition occurred 2 seconds later and lasted 6 minutes and 7 seconds, followed by the ignition of the S-IVB stage, giving a burn of 2 minutes and 25 seconds. This placed the S-IVB and the spacecraft into an Earth parking orbit with a perigee of 182km and an apogee of 187km. Three hours into the mission and after two orbits of the Earth, the S-IVB stage was fired to place the spacecraft into a simulated lunar trajectory. The burn lasted 5 minutes and when completed the spacecraft and the S-IVB were separated. The firing of the Service Module propulsion system, which placed the CSM into an apogee of 18,256km, with the heatshield of the Command Module (CM) away from the sun, followed this. The spacecraft was left in this position for 4.5 hours to test the thermal integrity of the spacecraft. Eight hours into the mission the CSM propulsion system was ignited once again and the spacecraft placed into a re-entry trajectory. The burn, which lasted 4.5 minutes, caused the spacecraft to enter the Earth's atmosphere at a velocity of 10.70km a second. All went well and the spacecraft was recovered from the Pacific Ocean off Hawaii by the aircraft carrier USS *Bennington*.

At 5.48 p.m. on 22 January 1968, an unmanned Saturn IVB rocket, Apollo V, blasted off from Launch Complex 37B at the Kennedy Space Center. Apollo V's mission was to verify the operation of the Lunar Module's structure itself and its two primary propulsion systems. It was also to evaluate the Lunar Module staging and the orbital performance of the S-IVB stage and instrument unit. Soon after achieving orbit, the nose cone, which replaced the CSM, was jettisoned and the Lunar Module separated. The first firing of the descent engine went virtually as planned, except that the Lunar Module's guidance system shut down after only 4 seconds of operation. This was because the engine's velocity did not build up as quickly as predicted. Houston pinpointed the problem in the guidance system itself and not in the hardware design. This enabled the engineers and scientists at Mission Control to pursue an alternative mission that achieved the same objective. After the mission had been completed at 2.45 a.m. the following day, the Lunar Module stages were left in decaying orbit to burn up on re-entry at a later date. A second unmanned flight had been scheduled, but due to the success of the first, Lunar Module II was cancelled. The LM II now stands in the National Air and Space Museum (NASM) in Washington DC.

Tragedy again struck the Russian space programme when on 27 March 1968 Colonel Yuri Gagarin, with his flight instructor Colonel Vladimir Serugin, took to the air from Chkalovsky, nr Star City, in a two-seater MiG-15UTI jet on a familiarisation flight. Gagarin was looking to qualify to fly the new MiG-17 fighter, his flying having gone by the board after his famous spaceflight. Later that morning two explosions were heard and the following afternoon the wreckage of the MiG-15UTI was found buried deep in the ground. Much speculation has been bandied about regarding the crash, even to the point of suggesting the Russian government had arranged it to rid themselves of Gagarin. But common sense prevailed, and the now accepted version is that the trainer went off course and into an area that was being used by supersonic fighters and was inadvertently 'jet washed' by one of them. This would have caused the trainer to go into a spin and consequently crash. All those involved in space exploration on both sides of the world mourned the passing of the world's first spaceman.

In the United States the second of the unmanned Apollo flights, Apollo VI, was launched from Complex 39A at the Kennedy Space Center on 4 April 1968. Two minutes and 13 seconds after launch problems were observed in the second stage during the boost phase. Four minutes later, two of the J-2 engines shut down early, so the firing of the remaining three engines was extended for 1 minute to compensate. This sent the third stage into a higher orbit than was planned, 177 x 363km rather than the 161km, near-circular orbit. The amount of fuel used forced the curtailment of the mission and the spacecraft was returned to Earth 9 hours and 50 minutes after launch, to be recovered by the USS *Okinawa*.

The time was now right for the first manned Apollo spaceflight. The crew had been training continuously since they had been told that they were to be the next prime crew. So too had their back-up crew of Frank Borman, Thomas Stafford and Michael Collins.

With thoughts of landing a man on the Moon firmly fixed in their minds, the Manned Spacecraft Center (MSC) directors had to look seriously at the spacecraft that would actually touch down on the Moon's surface. At Ellington Air Force Base, Texas, trials were being carried out on a Lunar Landing Research Vehicle (LLRV), but these received a setback on 6 May 1968 when NASA test pilot Neil Armstrong (Apollo XI) had to eject from the research vehicle after losing control. The LLRV was completely destroyed at a cost of $1.5 million.

On 11 October 1968, the first manned Apollo spacecraft, Apollo VII, lifted off from Cape Canaveral. Astronauts Walter Schirra (Commander), Don Eisele and Walter Cunningham had climbed into Apollo VII's Command and Service Module and blasted off into orbit around the Earth. They had trained to such an extent that they were probably the best-trained of all the astronauts to date, and it showed in their near flawless journey. Their flight evaluated all the major systems with the exception of the Lunar Module, and broadcast out the first live television commercial from space. The crew also drank water that had been produced

LLRV in flight.

as a by-product of the fuel cells. They carried out optical rendezvous experiments, platform realignment and sextant tracking of another vehicle.

One of the things that the crew discovered was to be important for future flights; this was the need to exercise and keep muscles in trim. Whilst they slept, they discovered that because of the lack of gravity their bodies invariably went into the foetal position; consequently they woke up with lower back and abdominal pains. The answer was to workout on a stretching device called an Exer-Genie.

There were also some personal problems that affected the flight. The commander, Walter Schirra, developed a very heavy cold and was, on occasion, extremely irritable with Mission Control, cutting off transmissions. The other two members of the crew also contracted the cold virus and, because of the weightless conditions, suffered a great deal more than they would have on Earth. The mucus that invariably accompanies a heavy cold can be easily discharged from the head when on Earth, but when in space it just fills the nasal passages and does not drain. The only means of relief was to blow hard to clear the passages, but this became very painful to the ears. This highlighted the problems that could be caused by a simple cold on spaceflights, especially if they were to be lengthy ones.

Because of the problem of nasal congestion, Schirra decided that the crew would not wear their suit helmets during re-entry, despite objections from Mission

Control. His decision was based on the fact that they would not be able to blow their noses during this period and the build up of pressure, were they to wear their helmets, could cause their eardrums to burst. The three crew members each took a decongestation tablet and re-entry was carried out without a problem.

After 164 orbits of the Earth the spacecraft re-entered the Earth's atmosphere on 22 October 1968, and landed 13km from the recovery ship, the aircraft carrier USS *Essex*. The splashdown in the Atlantic just south-east of Bermuda was the first time time flotation bags had been used. The moment the capsule hit the water it turned upside down, but immediately rolled upright when the flotation bags were inflated.

The NASA hierarchy did not like their decisions to be questioned or ignored, and choosing not to wear their helmets during re-entry against all advice was to guarantee that not one of the Apollo VII astronauts would be selected to fly in space again.

The launch of the RussianSoyuz 3 on 26 October was the first manned flight after the death of Komarov and was intended to carry out the first rendezvous and docking of two spacecraft. Soyuz 2, an unmanned spacecraft, had been launched earlier in the day, but the two failed to dock. The pilot of Soyuz 3 was Georgi Timo Feyevich Beregovi and the blame for the failure of the mission was placed upon his inability to manually control the spacecraft during the docking procedure. Ground controllers had manoeuvred the two spacecraft within 200m of each other before handing control over to Beregovi. His attempts at trying to dock with the other spacecraft caused him to use up almost all his fuel and the experiment was cancelled. Both spacecraft were safely returned to Earth three days later.

The Americans continued with their Apollo space programme with the launch of Apollo VIII (AS-503) on 21 December 1968, with astronauts Frank Borman, James A. Lovell Jr and William A. Anders on board. The spacecraft was placed into an Earth parking orbit whilst all the systems were re-checked, then S-IVB stage was re-ignited and the spacecraft was placed into a lunar trajectory. The crew then became the first men to leave the Earth's gravitational field.

On Christmas Eve that year the spacecraft's communications blacked out as it passed around the dark side of the Moon and the crew became the first to actually see the far side. On their way round, the crew took numerous photographs, and then as they came out of the communications blackout they carried out a live television broadcast to Earth. They read the first ten verses of Genesis and at the end wished all the viewers: 'Goodnight, Good luck, a Merry Christmas and God bless all of you – all of you on the good Earth.' It was said later that almost a billion people in sixty-four countries heard the astronauts reading from the Bible either on the radio or on television.

After spending Christmas Day orbiting the Moon the crew fired the CM propulsion system for 3 minutes and 24 seconds, having completed ten orbits of the Moon, placing the Apollo VIII spacecraft into an Earth trajectory and increasing its velocity to 3,875km/h. On reaching Earth orbit the Command Module (CM) separated from the Service Module (SM), re-entered the Earth's atmosphere and splashed down in the Pacific Ocean on 27 December, after 147 hours in space. The aircraft carrier USS *Yorktown* reached the recovery area, and helicopters from the ship recovered the crew and their spacecraft.

Amongst the numerous photographs taken by the crew were a number showing areas that were being considered for future landing sites.

Meanwhile the Russians, on 14 January, launched Soyuz 4 with cosmonaut Lieutenant General Vladimir Aleksandrovich Shatalov on board. The following day Soyuz 5 was launched with cosmonauts Colonel Yevegeniy Vasilyevich Khrunov, Colonel Boris Valentinovich Volynov and civilian engineer Aleksey Stanislovovich Yeliseyev aboard. The two spacecraft rendezvoused during the eighteenth orbit and completed the transfer of Khrunov and Yeliseyev from Soyuz 5 to Soyuz 4: the first time that crews had transferred from one spacecraft to another. This was an important breakthrough because one of the problems that had always concerned both cosmonauts and astronauts was how they would be rescued in the event of them being unable to return in their own spacecraft. The transfer of cosmonauts from one spacecraft to another proved that rescue was possible. Both spacecraft landed safely on 18 January.

Two months later, on 3 March 1969, the Americans launched Apollo IX (AS-504) with astronauts James A. McDivitt, David Scott and Russell Schweickart aboard. The launch had originally been scheduled for 28 February 1969, but was delayed to allow the crew to recover from mild respiratory illnesses. The normal launch phase was completed without incident and the S-IVB and Command and Service Module (CSM) placed into an Earth parking orbit of 192 x 189km. After all the systems had been re-checked, the CSM separated from the S-IVB, turned around and docked with the Lunar Module (LM) which was still attached to the S-IVB. The two spacecraft then separated from the S-IVB and were placed in an Earth-escape trajectory orbit.

James McDivitt and Russell Schweickart left the Command Module (Gumdrop), entered the Lunar Module (Spider) through the docking tunnel and carried out tests on the spacecraft's systems, performed a couple of telecasts and then fired its propulsion system. With all the tests completed, the two crew members returned to the CSM. On the following day they returned to the LM and carried out another telecast, before Russell Schweickart completed an Extra Vehicular Activity (EVA) in which he 'walked' between the two hatches of the spacecraft. During the EVA he took numerous photographs of both the two docked spacecraft, one of which showed David Scott standing in the open hatch of the CSM with the Earth behind him. He also commented on the rain squalls that were over the Kennedy Space Center at the time.

On 7 March, the two astronauts again entered the LM. This time David Scott separated the CSM from the LM and fired the Reaction Control System (RCS)

Command and Service Module (CSM) as seen from the LEM. Russell Schweickart seen here emerging from the Command Module to carry out a spacewalk.

thrusters, placing the two spacecraft 5.5km apart. At one point during the 6.5-hour separation the two spacecraft were 183km apart. With all the systems checked out and activities completed, the two astronauts docked with the CSM and returned to their spacecraft. The LM was then jettisoned. The remainder of the mission was spent taking multispectral photographs of the Earth and tracking NASA's meteoroid detection satellite, Pegasus III, which had been launched in 1965.

Apollo IX re-entered the Earth's atmosphere on 13 March and splashed down in the Atlantic east of the Bahamas. The aircraft carrier USS *Guadalcanal* recovered the spacecraft and its crew within 1 hour.

On 18 May Apollo X (AS-505) was launched, with astronauts Thomas Stafford, Eugene Cernan and John Young aboard. After going into an Earth parking orbit and re-checking the systems, the S-IVB engine was ignited and the spacecraft placed into a lunar trajectory. One hour later, whilst on course for the Moon, the

Command and Service Module (CSM) separated from the S-IVB, transposed and docked with the Lunar Module (LM). The S-IVB was then placed into a decaying solar orbit.

After entering lunar orbit, Stafford and Cernan transferred to the Lunar Module (LM), undocked from the CSM and headed toward the lunar surface. They descended to within 14,300m of the surface and then returned to dock with the CSM. The LM was then placed into a solar orbit and the crew prepared to return to Earth. After entering Earth orbit, the SM was jettisoned and the CM re-entered the Earth's atmosphere. Their spacecraft was recovered safely by the USS *Princeton* after splashing down in the Pacific Ocean. The stage was now set for the greatest exploration mission of all time – a visit to another world.

On 16 July 1969 Apollo XI (AS-506) lifted off the pad at Cape Kennedy (the former name of Cape Canaveral), with astronauts Neil Armstrong (Commander), Edwin 'Buzz' Aldrin Jr (LEM Pilot) and Michael Collins (CSM Pilot) aboard. Four days later Neil Armstrong and Buzz Aldrin transferred to the Lunar Excursion Module (LEM) (Eagle) from the Command and Service Module (CSM) (Columbia) and after checking that all the systems were working properly, they separated from the CSM and headed toward the lunar surface and into the history books.

Sixty miles above the surface of the back side of the Moon, the on-board computers made the first burn to slow the LEM down. The burn lasted for 28.5 seconds and the spacecraft coasted towards the front side of the Moon. Inside the LEM Armstrong and Aldrin carried out detailed checks of the on-board systems as they rapidly approached the moment when the descent engine would fire up and take them on to the surface. Right on time the descent engine fired, then 26 seconds later the engine went to full power, rapidly de-accelerating the spacecraft. Whilst this was happening the two astronauts watched the primary and secondary computers intensely. Buzz Aldrin carried out a running commentary to Neil Armstrong, comparing the read-out sequences with their information cards, and all the time the same information was being transmitted back to Mission Control.

As they continued their descent, the two astronauts suddenly became aware of a yellow programme alarm, in the shape of a caution light appearing on the computer. They asked the computer to define the problem and the reply was that it was being overloaded with questions and being given too little time to answer them, therefore it could not cope. A second alarm light then appeared and all the time the LEM was rapidly descending towards the surface of the Moon. In the spacecraft and back at Houston, everyone waited with baited breath to see what was going to happen next and whether or not the landing would have to be aborted. Back in Mission Control, the man responsible for the computers in the LEM, Steve Bales, cut in and told the CapCom, Charlie Duke, to tell the two astronauts to ignore the computers and go for landing. Two more warning lights

Portrait photograph of the crew of Apollo XI: Neil Armstrong, Mike Collins and Buzz Aldrin.

suddenly appeared and once again Bales told the crew to ignore the computers.

It was discovered later that the computer programmers from MIT, who had designed the on-board computer landing programme in the LEM that interrogates the landing radar, had never spoken to the computer programmers who designed the rendezvous landing radar programme. It is interesting to note that today's mobile phones have more power than the computers aboard that Apollo IX spacecraft.

At 500ft Armstrong took over manual control, and the two astronauts peered out of their observation windows at the rapidly approaching surface. They saw the area they were approaching was strewn with rocks and boulders, but noticed that the area beyond was clear. Neil Armstrong immediately extended the LEM's trajectory and touched down in the clear area. Whilst this was happening, Mission Control was monitoring the fuel rate and warned Armstrong that he had only 60 seconds of fuel left. Then came the warning '30 seconds', '10 seconds' and then touchdown.

As the spacecraft touched down in the Sea of Tranquility, Buzz Aldrin shut down the descent engine. Then, from Neil Armstrong, the words that the whole world was waiting to hear came over the loudspeakers in Mission Control at Houston:

Houston, Tranquility Base here. The Eagle has landed.

The relief both in the LEM and in Mission Control was enormous and for a few minutes a feeling of complete exhaustion swept through them all, before it gave way to immense apprehension and excitement. In the LEM, frantic preparations started for the two astronauts to step out on to the surface of the Moon.

The following day Neil Armstrong climbed down the ladder of the Lunar Module and stepped out onto the Moon's surface, the first man to set foot on another world, and spoke the now immortal words:

That's one small step for ah … man – one giant leap for mankind.

LEM Eagle heading towards the surface of the Moon.

Shot showing the desolate landscape of the Moon as seen for the first time by man.

When Neil Armstrong went to place his foot on the Moon's surface, he did it very carefully because the shock absorbers on the legs of the LEM had not deployed. This left a gap of 3.5ft between the bottom rung and the Moon's surface. He also had to make sure that once he had stepped on the Moon he could get his foot back on the rung of the ladder.

The egressing of the LEM was not without its problems. Armstrong found that the hatch was extremely difficult to get through with the Portable Life Support System (PLSS) on his back, and the Remote Control Unit (RCU) on his chest prevented him from seeing his feet. This meant that the climb down the ladder to the surface was by feel only. When Buzz Aldrin exited the LEM he had to make sure that he did not close the hatch behind him because there was no handle on the outside. What would have happened had he closed the door one can only speculate!

Twenty minutes later, Buzz Aldrin stepped down from the ladder to join Armstrong on the Moon's surface. The two men gazed around for a few minutes taking in the breathtaking stillness of the lifeless world, then got to work deploying experimental instruments and collecting samples. The following day, with all the samples collected, the Lunar Module blasted off the Moon's surface and into a lunar orbit to rendezvous with Michael Collins in the CSM. Before leaving the Moon, Neil Armstrong placed beside the lower section of the Lunar Module two Russian space medals belonging to Yuri Alexseyevich Gagarin and Vladimir Mikhaylovich Komarov. The medals had been given to American astronaut Frank

Borman, Commander of Apollo VIII, by the cosmonauts' respective wives whilst he was on a goodwill tour of the Soviet Union, requesting that they be placed on the Moon when the Americans eventually got there.

As Buzz Aldrin entered the Lunar Module to prepare to blast off the lunar surface, his backpack caught and snapped the firing switch lever to fire their engines. Fortunately he managed to stick a pen into the switch and connect the contacts. It worked, and saved the Moon mission from certain disaster.

After reaching Moon orbit the LEM rendezvoused with the CSM. After a jubilant meeting with Michael Collins in the CM they transferred all the samples of rock to the CSM and sent the LEM crashing down on the surface of the Moon so that seismologists back on Earth could monitor the effect. It has to be remembered that Mike Collins was alone in the CSM whilst the other two astronauts were attracting all the attention on the Moon. He was also carrying out numerous tasks and had to be in the right place to link up with the upper stage of the LEM when it blasted off the Moon's surface.

Setting a course for Earth, the crew settled down to reflect on what they had achieved. Just before reaching orbit they jettisoned the Service Module and prepared for re-entry. The crew and their Apollo XI spacecraft splashed down in the Pacific Ocean on 24 July. Their lives would never be the same again. The three men had been trained for every aspect and eventuality of the flight to the

Buzz Aldrin standing on the surface of the Moon. Neil Armstrong's reflection can be seen in his visor.

Another photograph of the Moon's desolate surface, this time with the LEM in the background.

Moon, but they were not trained for the adulation they would receive when they returned home.

On returning to Earth Buzz Aldrin submitted a travel claim form.

The Russians, now realising that the Americans had won the race to put a man on the Moon, concentrated their efforts on carrying out rendezvous experiments, with future thoughts of putting a Space Station into orbit. On 11 October 1969, they launched Soyuz 6 with cosmonauts Georgi Stepanovich Shonin and Valeri Nikolayevich Kubasov aboard. The following day Soyuz 7 was launched with cosmonauts Vladislav Nikolayevich Volkov, Anatoli Valisyevich Filipchenko and Viktor Vasilyevich Gorbatko aboard. They carried out rendezvous manoeuvres whilst in orbit round the Earth and then were joined by Soyuz 8, which was launched on 13 October with cosmonauts Vladimir Aleksandrovich Shatlov and Aleksey Stanislovovich Yelisyev aboard. The three spacecraft carried out a variety of experiments all centred around rendezvous techniques, but because of a communications failure in the radio locations system, Soyuz 7 and 8 never got closer than 1,600ft. The whole exercise was to have been filmed by Soyuz 6.

Meanwhile, in the United States, the Americans were preparing to carry out a second lunar landing. On 14 November 1969 Apollo XII (AS-507) blasted off from the pad at Cape Kennedy with astronauts Charles 'Pete' Conrad Jr, Alan L. Bean and Richard F. Gordon Jr aboard. Thirty-six seconds after lift-off the whole spacecraft experienced a total loss of electrical power after being struck by lightning and again at 52 seconds, but power was restored and the spacecraft headed toward an Earth parking orbit.

After carrying out all the necessary re-checking that accompanied every spaceflight, the S-IVB propulsion system was fired up and the spacecraft placed on a lunar trajectory. One hour into the flight Pete Conrad and Alan Bean transferred to the LM and then proceeded with Richard Gordon, to separate from the S-IVB and transpose with the CSM (Yankee Clipper). Once this had been achieved the S-IVB was sent on its way into a solar orbit, and the CSM with the LM attached went on its way to the Moon. During the transposing the LM crew carried out a live television broadcast showing colour pictures of the actual transposing, the interior of the LM and views of the Earth and Moon as the two spacecraft raced towards it.

After entering orbit around the Moon, Conrad and Bean undocked the LEM (Lunar Excursion Module) from the CSM and headed towards the lunar surface. The spacecraft touched down in the Ocean of Storms just 180m from the Surveyor spacecraft that had landed there three years earlier. Conrad was some inches shorter than Neil Armstrong, and as he negotiated the last step on the LEM ladder he exclaimed as he touched the surface:

> Whoopee! Man, that may have been a small step for Neil, but that's a long one for me.

The LEM had landed 163m from the unmanned Surveyor III spacecraft that had landed on the Moon's Ocean of Storms in 1967. The two astronauts carried out two EVAs in which they deployed ALSEP (Apollo Lunar Surface Experiments Package). They also recovered parts from the Surveyor III, including its television camera and soil scoop. Above them, orbiting the Moon in the Apollo XII spacecraft, Richard Gordon carried out a lunar multispectral photography experiment, photographing future landing sites.

After 31 hours and 31 minutes the two astronauts lifted off the lunar surface to rendezvous with the CSM. The LEM (Intrepid) was jettisoned and sent crashing back on to the surface of the Moon, where the reverberations, which lasted 30 minutes, were recorded by a seismometer. The crew landed safely in the Pacific Ocean on 24 November.

On 11 April 1970 Apollo XIII (AS-508) was launched from Cape Canaveral for the third planned landing on the Moon, with astronauts Jim Lovell, Jack Swigert and Fred Haise on board. Events soon took a turn for the dramatic.

During launch an incident occurred that almost resulted in a launch abort. The second stage engine started to oscillate violently, luckily causing it to shut down

Apollo XIII patch.

early. The engine, which weighed 2 tons, was bolted to the massive thrust frame and was bouncing up and down at 68G. This caused the frame to flex 3in (76mm) at 16 hertz! Fortunately, after 3 seconds of these oscillations, the engine's 'low chamber pressure' switch tripped. Fortunate was the right word, as the switch had not been designed to trip in this manner – but it did and the engines automatically shut down. If the oscillations had continued, there is no doubt that the Saturn V rocket would have been torn apart.

With the problem resolved, the launch went ahead without further incident. After entering an Earth parking orbit all the systems were checked out before the S-IVB stage sent the spacecraft on a translunar trajectory. The separation from the S-IVB and the transposition of the CSM (Odyssey) went according to plan and was televised. Everything went perfectly until just after 56 hours into the flight, when Mission Control asked Swigert to switch on the fans in oxygen tanks 1 and 2 in the SM and stir up the oxygen. Minutes later there was a loud bang.

Jack Swigert's apprehensive voice came over the radio:

> 'OK Houston we've had a problem here.'
> 'Say again?'
> 'Houston we've had a problem.'

This was the first indication to Mission Control that there was something seriously wrong. It was discovered later that one of the two oxygen tanks in the Service Module (SM), had ruptured because one of the thermostatic safety switches in the tank had been welded shut by an electrical arc when a 65-volt DC power was applied through a switch designed to take only 28 volts DC. It was then thought that this probably caused the heater tube assembly to overheat, which in turn severely damaged the Teflon insulation on the fan motor wires. This, in all probability, short-circuited and ignited the insulation and a portion of the tank. The rapid expulsion of high-pressure oxygen that followed would have blown off the outer panel to bay 4 of the SM, causing a leak in oxygen tank 2.

The pressure in oxygen tank 2 increased rapidly until the top blew off. The insulation between the inner and outer shells of the tank then caught fire and within seconds the whole of the bay was turned into a raging inferno when the Mylar insulation that lined the bay also ignited. The gases building inside the bay reached bursting point, and seconds later the cover of the bay was blown off and Apollo XIII was in trouble. The one saving grace was that the pressure in the oxygen tank had blown the top off. Had it continued to increase without releasing, it might have blown the Command Module (CM) off the SM like a cork being shot out of a bottle.

Inside the CM the crew was unaware of what was happening behind them. They had felt a shudder pass through the spacecraft accompanied by a small bang, but because they were out in space, there was no air to carry the sound or shock waves of the explosion. The increase of lights flashing on the telemetry screens and the sudden loss of power in the spacecraft made the crew aware that something was seriously amiss. They were unaware of what had happened, but two of the three fuel cells had gone dead and the remaining one was beginning to drain slowly.

There was no way of repairing the damaged tank and no way of turning the spacecraft round and heading back to Earth. The crew had to go on and they were already 200,000 miles from Earth. A Moon landing was out of the question; it was now a matter of survival. The three-man crew moved into the Lunar Module (Aquarius) which was only ever designed to accommodate two crew members. It soon became obvious that the crew's oxygen supply was rapidly being depleted, so the LEM's life support systems had to be modified to support the three astronauts. The spacecraft was now operating on battery power, so every piece of non-essential equipment was switched off, including the heating.

The LEM was now the prime spacecraft and was still attached to the now lifeless CSM. Never designed to be used to travel through space towing another section of the spacecraft, the LEM had to undergo manual mid-course corrections so that the crew could place it into an orbit of the Moon that would place them on a trajectory back to Earth.

The carbon dioxide filter on Apollo XIII.

With the air inside the LEM becoming more and more toxic, the ground support team devised a carbon dioxide filter using the only things that would be available in the spacecraft: cardboard, plastic bags and tape. Within an hour of the crew making a couple of filters the level of carbon dioxide had dropped dramatically.

After a fraught journey, and thanks to the ingenuity of the engineers back at Houston, the crew returned safely to Earth. One person who deserved praise was astronaut Ken Mattingly. A member of the original crew, Mattingly was bumped off the mission because he had been in contact with someone who had measles. Mattingly placed himself in the simulator attempting to emulate the same conditions as those being experienced by his fellow astronauts aboard Apollo XIII, even down to using a torch identical to the one the crew were using. When different ideas were broached to help them survive he acted them out in the cramped confines of the Lunar Module simulator to see if they would work. Ken Mattingly spent almost the same length of time in the simulator as did the Apollo XIII astronauts in space.

As the spacecraft approached Earth, the crew powered up the CM and left the LEM, their 'lifeboat', with the words 'Goodbye Aquarius and Thank You'. As their spacecraft splashed down in the Pacific Ocean, the aircraft carrier USS *Iwo Jima* was standing close by and in less than an hour the three astronauts were aboard ship. The greatest rescue mission of all time had been accomplished. There was one sobering thought: if the explosion had happened after the crew had visited the Moon, there was no way they would have survived. Their 'lifeboat' would have been left behind on the Moon's surface!

It is interesting to note that despite the Cold War relations that existed between the USSR and the USA at the time, the Soviets, on hearing of the plight of the three astronauts, immediately offered their help unreservedly, both at sea and in the air in the event of the spacecraft going off course after re-entry and coming down in their part of the world.

Later the Grumman Company, who had made the Lunar Module 'Lifeboat', sent Rockwell (who had made the Command and Service Modules), a light-hearted invoice for $312,421.24 for towing the damaged module back to Earth.

Rockwell replied in a similar vein, issuing a statement saying that before Grumman embarked upon such a claim, it should remember that North American Rockwell had not received payment for ferrying Lunar Modules on previous trips to the Moon.

The Russians continued to carry out space duration tests and on 2 June 1970 Soyuz 9, with cosmonauts Major-General Andriyan Grigoryevich Nikolayev and civilian engineer Vitali Ivanovich Sevastyanov aboard, carried out the longest endurance record for a manned spaceflight of 17 days, 16 hours and 59 minutes. The Russians were slowly accumulating great expertise in time spent in space and the effects of weightlessness on the crews. The length of time the two

GRUMMAN AIRCRAFT ENGINEERING CORPORATION
PURCHASE REQUISITION

A374979

NOTE: SEE CORPORATE PROCEDURE M110 BEFORE COMPLETING THIS FORM

DO NOT WRITE IN THIS AREA — FOR PURCHASING USE ONLY

SELLER INVITED TO QUOTE		BUYER	CODE	TELEPHONE - AREA CODE 516		DATE REC'D IN PURCH.
1) North American Rock	4)	North American Rock		LR 5-		
2) Pratt & Whitney	5)	CONTRACT NO. UR B(00) B(00)		CERTIFIED FOR NATIONAL DEFENSE USE UNDER ... ARTICLE ... ON BACK.	PRIORITY 1	DATE RELEASED TO TYPE 4/13/70
3) Beech Aircraft	6)	SUBJECT TO GOVERNMENT INSPECTION AT [X] YOUR PLANT [ ] GAEC [ ] NONE		FOR DESTINATION (UNLESS OTHERWISE INDICATED)		TERMS Cash
SELLER AWARDED PURCHASE ORDER		PURCHASE ORDER NUMBER		DATE		

SHIP TO: Hou - MSC

VIA: LM-7, USS Iwo Jima, GOVAIR

DELIVERY REQUIRED AT GAEC — DATE: None

SELLER PROMISE: Never Again

ITEM NO.	QUANTITY	UNIT	PART NO. — DESCRIPTION	I-T	ACCT. NO./JOB NO.	TAX CODE	UNIT PRICE
1	400,001	Mi	Towing, $4.00 first mile, $1.00 each additional mile Trouble call, fast service				$400,004.00
2	1	KWH	Battery Charge (road call - $.05 KWH) customer's jump cables				4.05
3	50#	#	Ox Oxygen at $10.00 /lb				500.00
4	1		sleeping accomodations for 2, no TV, air-conditioned, with radio, modified american plan, with view		NAS-9-1100		Prepaid
5			Additional guest in room at $8.00/night (1) Check out no later than noon Fri. 4/17/70, accommodations not guerenteed beyond that time				32.00
6			Water				No Charge
7			Personilized "trip -tik", including all transfers, baggage handling and gratuities				No Charge
			Sub Total				$400,540.05
			20% commercial discount + 2% cash discount (net 30 days)			(-)	83,118.81
			total				$312,421.24
			No taxes applicable (government contract)				

SUGGESTED SOURCES/REMARKS (INCLUDE CWA NO. IF APPLICABLE):

RECEIVING DELIVER TO: USS Iwo Jima — VIA: Air Express

REQUESTED BY: NASA (MSC) — EXT. — PLT. & DEPT. NO. — DATE

APPROVED BY — DATE

Grumman's invoice for towing Rockwell's CSM - Apollo 13.

The invoice sent to Rockwell by Grumman for 'towing' the crippled CSM back to Earth.

Soyuz 9 cosmonauts spent in space was significant inasmuch as they suffered quite badly from the effects of living in zero gravity. For over two weeks they had difficulty in walking, stressing the necessity for cosmonauts/astronauts to carry out strict and rigorous exercise whilst in space.

Then, on 12 September 1970, the Russians launched Luna 16, an unmanned spacecraft. The mission was to land on the Moon and bring back samples from the Sea of Fertility. The mission was a complete success and Luna 16 was the first unmanned spacecraft to land on the Moon and return. One month later Luna 17 was launched on another unmanned mission to the Moon. After a soft landing in the Sea of Rains the spacecraft released Lunokhod I, a self-propelled vehicle resembling a large bathtub on wheels. The vehicle, about the same size as a Volkswagen Beetle, was propelled on eight spoked wheels, powered by solar energy that in turn powered batteries. It was equipped with scientific apparatus, instruments, television cameras and a radio transmitter and receiver. Despite its ungainly appearance it was extremely successful and in its operational life, logged 8,458 miles and explored 400,000 square miles.

The Americans responded to the Russian achievements with the launch, on 31 January 1971, of Apollo XIV (AS-509) with astronauts Alan B. Shepard Jr, Stuart Roosa and Edgar D. Mitchell aboard. This was the most inexperienced crew to

ALSEP being deployed by Alan Shephard.

go into space as only Alan Shepard had been before and that had been for just 15 minutes. As it turned out this was to be one of the most successful Apollo missions, and was almost trouble-free.

On reaching Earth parking orbit, as with all previous missions, the astronauts checked out the systems before firing up the S-IVB and putting the spacecraft into a translunar trajectory. During the transposition of the LEM (Antares) and the CSM (Kitty Hawk), the crew experienced difficulty in docking. By the fifth attempt considerations were being given to a possible abort, but after the sixth attempt the latches activated and the two spacecraft were mated. An inspection of the latching mechanism was carried out, but nothing could be found and the cause of the difficulties could not be determined. On entering lunar orbit on 5 February 1971 Shepard and Mitchell entered the LEM and headed down to the lunar surface.

After separating from the Command Module in lunar orbit, the LEM also had two serious problems. First, the LEM computer began getting a signal from a faulty switch. When the switch was simply tapped on the fault cleared, but if the problem recurred after the descent engine fired, the computer would think the signal was real and would initiate an auto-abort, causing the ascent stage to separate from the descent stage and climb back into orbit. NASA and the software teams at MIT scrambled to find a solution, and determined the fix would involve re-programming the computer to ignore the false signal. The software modifications were transmitted to the crew via voice communication, and Mitchell manually entered the changes just in time.

A second problem occurred during the powered descent, when the LEM radar altimeter failed to lock automatically onto the Moon's surface. This prevented the navigation computer from obtaining the information on the vehicle altitude and groundspeed. The astronauts cycled the landing radar breaker and a signal near 50,000ft (15,000m) was obtained, again just in the nick of time. Shepard then manually landed the LEM close to its intended target. Mitchell believes that Shepard would have continued with the landing attempt without the radar, using the LEM's inertial guidance system and visual cues. But a post-flight briefing of the descent data showed the inertial system alone would not have been enough, and the astronauts probably would have been forced to abort the landing as they approached the surface.

After landing in the Fra Mauro area the two astronauts carried out two EVAs, deployed ALSEP and used the first Mobile Equipment Transporter (MET) to carry 43kg from a variety of areas. The crew also brought back the largest payload from the Moon and took the first colour television pictures.

Apollo XIV crew being recovered from their capsule.

Just before leaving the Moon Alan Shepard dropped a golf ball onto the surface and, on the third swing, drove the ball a distance of 366m. Having established the most exclusive golf club in the universe, the two astronauts returned to the spacecraft. After stowing away 43kg of lunar rock and dust they lifted off the following day and rendezvoused with the CSM for the return to Earth.

After transferring all the samples and equipment from the LEM to the CSM the ascent stage of the LEM was jettisoned and sent crashing back into the Moon's surface, where the impact was recorded by the ALSEPs left by Apollo XII and XIV. After an uneventful voyage home, the Apollo XIV spacecraft landed on 9 February in the Pacific Ocean.

The crew of Apollo XIV was the last to be placed in quarantine as tests on the crew of Apollo XI proved to be negative for any alien virus.

The Russians took the pace of the space race up a gear, when, on 19 April 1971, they launched Salyut 1, the first manned Space Station. It was to spend nearly six months in orbit before burning up on re-entry. The first cosmonauts to visit the station were aboard Soyuz 10 and were Lieutenant-General Vladimir Aleksandrovich Shatalov, civilian engineer Aleksey Stanislovovich Yeliseyev and civilian engineer Nikolay Nikolayevich Rukavishnikov. Their spacecraft took off on 23 April 1971 and docked with the Space Station for 5.5 hours, but did not transfer any of the crew. Officially no reason was given for the short duration of the mission, but it was thought that the transfer

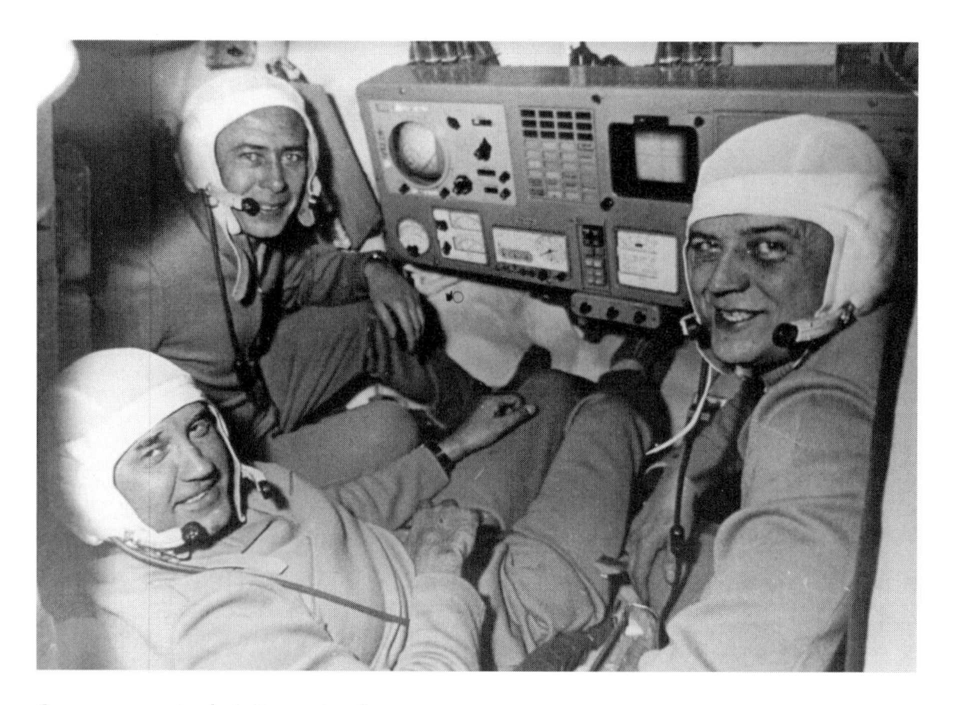

Soyuz 11 crew in their Soyuz simulator.

Attempts to resuscitate the crew of Soyuz 11 after they had been discovered unconscious in the spacecraft.

tunnel had been damaged. It wasn't until 6 June that the first cosmonauts went aboard Salyut.

During the rendezvous the air inside the Soyuz 10 spacecraft became toxic and Rukavishnikov lost consciousness. He was revived and the spacecraft separated from the Space Station and returned to Earth.

Death again marred the Russian space programme when, after a seemingly successful mission, the crew of Soyuz 11 died. Soyuz 11 was launched with cosmonauts Georgy T. Dobrovolsky, civilian engineer Vladislav Nikolayevich Volkov and civilian engineer Viktor Ivanovich Patsayev aboard. After spending over three weeks in the orbiting Space Station and carrying out a number of medical experiments, investigations and astronomical observations using the Orion astro-physical observatory, the crew returned to their spacecraft and headed back to Earth.

The mission itself hadn't been without incident. Two weeks into the flight a cable caught fire and it was with great difficulty that Chief Designer Mishin was able to calm the three cosmonauts down and persuade them to continue. There were a number of other problems that caused the mission to be terminated a week earlier than had been planned.

As the spacecraft entered the Earth's atmosphere everything was confirmed with the three cosmonauts that all was well, and then they entered into the communication blackspot. The parachutes deployed automatically and landed the spacecraft safely, but when the capsule was opened the crew was found to be dead. It was discovered later that a ball joint in an exhaust valve had been accidentally dislodged, decompressing the capsule. Despite desperate attempts by Patsayev, who tried to plug the leak with his finger, the air in the capsule had leaked out. None of the cosmonauts were wearing spacesuits, which would have saved their lives. This was a dreadful blow to the Russian space programme and it was to be more than two years before any of their cosmonauts entered space again. Chief Designer Mishin later stated that the crew could have used a manual drive to close the valve, but either they forgot or had never been shown how to employ it.

America continued to carry out missions to the Moon, this time aboard Apollo XV (AS-510). The spacecraft was launched on 26 July 1971, with astronauts David R. Scott, James B. Irwin and Alfred M. Worden aboard. As with all previous missions, the spacecraft went into an Earth parking orbit just 12 minutes after lift-off. When all the checks had been completed, the S-IVB propulsion system was fired up and the spacecraft was sent into a translunar trajectory.

During the flight to the Moon, the transposition of the LEM (Falcon) from the S-IVB and the mating with the CSM (Endeavour) took place. Scott and Irwin entered the LEM after going into lunar orbit and headed for the surface of the Moon. The spacecraft landed in the Hadley-Apennine region on 30 June. The two astronauts carried out five EVAs and the first real exploration by means of the Lunar Roving Vehicle (LRV).

The LRV gave the appearance of a chassis with a wheel in each corner and two seats perched at the rear. The tyres were made out of piano wire that had been stretched to form a surface similar to that of a rubber tyre. It was powered by sealed electric motors in the wheel hubs and ran on two 36-volt batteries. Mounted on the buggy was a complete communications package that kept the astronauts in radio and televisual contact with Mission Control. The whole LRV weighed a staggering 455lb on Earth, but on the Moon weighed only 76lb, and the cost and development of this little four-wheel drive, off-road dune buggy was a mere $8 million. It could carry 2.5 times its weight at a speed of 10mph. It was a weird, flimsy-looking machine but it worked, and it enabled the astronauts to explore farther away from their spacecraft and obtain samples of rock from various locations. One of the rocks, named the Genesis Rock, was age-dated at 4.15 billion years old, give or take 25 million years, by the University of New York. The oldest thing ever found on Earth was dated at 3.3 billion years, so another piece of puzzle to the creation of the universe was found.

The LRV was stripped of everything that could be taken back to Earth, with the exception of the camera. That was pointed at the LEM and recorded the ascent stage blasting off the surface of the Moon. Just before the two astronauts entered the LEM for the last time, they placed a plaque on the surface of the Moon inscribed with the names of all the astronauts, both American and Russian, who had died in the pursuit of space exploration. As the CSM was about to leave lunar orbit, the crew launched a sub-satellite to carry out scientific studies of the Moon from a low orbit.

On the return journey Command Module pilot Al Worden carried out an EVA to recover some of the scientific instruments mounted on the outside of the spacecraft. The CM returned safely to Earth on 7 August 1971 and was recovered

The crew of Apollo XV: Dave Scott, Al Worden and Jim Irwin.

Apollo XVI lifting off from the Cape.

by the aircraft carrier USS *Okinawa*. Another successful trip to the Moon had been accomplished.

The Russians launched another unmanned lunar probe, Luna 18, on 2 September 1971. It attempted to land near the Sea of Fertility, but on landing all communication with the probe ceased. It is believed that the probe crashed into the Moon's surface and was destroyed. Not to be discouraged, another probe was launched on 28 September, Luna 19. This landing was successful and it carried out geophysical research of the Moon's gravitational field and sent back more photographs of the surface.

Then on 14 February 1972 Luna 20 was launched. This was one of the most ambitious projects carried out by the Russians as the unmanned spacecraft was designed to land on the Moon, retrieve some samples and return them to Earth. The spacecraft landed on the Moon between the Sea of Fertility and the Sea of Crisis on 21 February. An Earth-operated drilling rig started to drill into the lunar surface to a depth of 35cm. Samples were obtained and transferred to a container that was hermetically sealed and returned to Earth. The result of the analysis of the samples showed that the area consisted of anorthosite, which contrasted greatly with samples obtained by Luna 16 from the Sea of Fertility, which was primarily basaltic rock.

The sixth manned mission to the Moon, Apollo XVI (AS-511), was launched on 16 April 1972. On board were astronauts John W. Young, Thomas L. Mattingly II and Charles M. Duke Jr. On reaching Earth orbit the crew put the spacecraft into a holding orbit while they checked out the systems. This done, the S-IVB and the spacecraft were put into a translunar trajectory. About 1 hour into the flight the CSM (Casper) jettisoned the S-IVB and transposed with the LEM (Orion). On entering lunar orbit John Young and Charles Duke entered the Lunar Module and powered up all the systems. On separation the two spacecraft stayed on station whilst the crew evaluated the service propulsion system. There had been a nagging doubt about the propulsion system, but it was decided that it was safe and could be used if required.

After orbiting the Moon John Young and Charles Duke landed the LEM in the Descartes region and decided to take a rest break before carrying out their first EVA. The following morning they unpacked the LRV and prepared to carry out an exploration in the region of Survey Ridge, Stone Mountain and North Ray Crater areas.

Three EVAs were carried out all using the LRV, and a number of investigations and experiments were completed. They also transmitted live colour television pictures back to Earth and, when they departed, their launch from the surface of the Moon was captured live on television from the camera mounted on the LRV. After transferring to the CSM the LEM was jettisoned and, like the other LEMs before it, meant to impact on the Moon's surface. This was so that seismic measurements could take place, enabling scientists back on Earth to try and measure

Ron Evans with the Lunar Rover in the Mare Serenitatis are of the Moon.

the depth of the Moon's crust. Unfortunately the LEM went into a decaying lunar orbit around the Moon. As the Apollo XVI spacecraft left lunar orbit it deployed a scientific sub-satellite, but that too ran into problems and like the LEM went into a decaying orbit.

Whilst in orbit around the Moon, Ken Mattingly became aware of flashes of light coming from the Moon's surface. It was never discovered what they were but suffice to say the UFO theory was again raised. The Apollo XVI spacecraft returned to Earth safely and splashed down in the Pacific Ocean on 27 April. It was recovered by the aircraft carrier USS *Ticonderoga*.

The final manned Apollo lunar explorer, Apollo XVII, was launched on 7 December 1972 at 0033 hours EDT from the Kennedy Space Center. The crew, Eugene Cernan, Ronald Evans and Harrison Schmitt, had to spend an additional 3 hours in the CSM because of a countdown sequence failure. This was the only hardware failure during the entire Apollo programme to cause a launch delay. Earth orbit was achieved and the insertion into a lunar trajectory carried out without problem. The transposition of the CSM with the LEM was completed during the translunar flight and again was incident free. A number of scientific experiments were carried out during this part of the flight to the Moon as on all previous missions. Apollo XVII entered lunar orbit on 11 December and after completing the separation of the LEM from the CSM Cernan and Schmitt made

Apollo XVII Lunar Rover on the Moon.

their descent to the surface. The first EVA began 4 hours later and, after off-loading the LRV and experimental packages, the two astronauts decided to go ahead with some of their tasks rather than wait until the following day.

The area in which the LEM had landed was known as Taurus-Littrow and was near the coast of the great frozen sea of basalt, the Mare Serenitatis. The unique visual beauty of this valley was the epitome of an ethereal vision. Of all the landing sites chosen, it would have been hard to find one that ended the exploration of the Moon in such a memorable way. The second EVA lasted 7 hours and 37 minutes and it was during this period that the now-famous orange soil was discovered. It was to be the subject of geological discussion for many years to come.

The third and final EVA was on 13 December, during which a great variety of geological samples were taken. Just prior to entering the LEM for the last time, a plaque was unveiled on the landing gear. It said quite simply: 'Here man completed his first exploration of the Moon, December 1972 AD. May the spirit of peace in which we came be reflected in the lives of all mankind.'

It was signed by the crew of Apollo XVII and by the then President of the United States of America, Richard M. Nixon. The last crew then left the Moon and returned to Earth knowing it would be years before man ever landed there again – if ever.

CHAPTER FIVE

A NEW ERA

With the end of the Apollo space programme, Grumman's president, Lew Evans, directed that the company should concentrate its efforts on the development of a Space Shuttle-type vehicle. In 1972 President Nixon had announced that the United States was going to build a Space Shuttle Transportation System, a combination of spacecraft and aircraft that was capable of operating in an Earth orbit. The launch costs would be drastically reduced from those of the Mercury, Gemini and Apollo programmes. The shuttle would fit the needs of many types of mission and would reduce the costs of payloads, through the elimination of current size, weight and reliability constraints of the one-shot launches. It would not only be able to deliver civilian and military satellites into space, but would also provide the means by which they could be maintained or even returned to Earth for repair. In addition to normal astronaut crews it would also carry non-astronaut passengers interested in conducting scientific and other investigations. Moreover, it could be used as a rescue vehicle in the event of an accident in space involving another manned spacecraft.

Back in the 1960s, atmosphere glide tests at hypersonic speeds were carried out by reliable, non-ablative, lifting bodies. Lifting bodies, or to give them their full title, lifting-body research vehicles, were the brainchild of Dr Alfred J. Eggers Jr. The whole idea of the aircraft/spacecraft concept was that the craft could attain aerodynamic stability and lift without the use of wings from a specially shaped body. The craft would also be capable of landing on a runway at normal speeds. The results from these experimental craft confirmed the feasibility of controlling lifting bodies during critical phases. The idea of a Space Shuttle grew from these tests, together with the mounting pressure to reduce the cost of placing payloads into Earth orbit by means of disposable rockets.

The space programme in America suddenly went into a lull, just as if someone had questioned what might come next. A great deal of thought had been given to the waste of materials that constituted the launch of the Mercury, Gemini and Apollo spacecraft. From the massive rocket that left the launch pad, all that returned to Earth was the tiny little capsule that had been perched on top. The remainder either burnt up on re-entry or became part of the increasing junkyard that circled the Earth.

The Russians, on the other hand, were still in experimental mode, and the development of the Salyut Space Station and its obvious potential caused the Americans to think seriously about developing and constructing their own. Having said this, the launch of Salyut 2 on 3 April 1973 was to be another setback in the Russian space programme. Tracking stations from around the world started detecting anomalies coming from the area where the Salyut Space Station was supposed to be settling in orbit. It was quite obvious to those tracking the Space Station that it was in trouble and coming apart. The Russians denied that there was a problem and said that Salyut 2 was settling into its pre-arranged orbit.

On 25 April the tracking stations in North America announced that the Salyut 2 Space Station was breaking up, despite what the Russians were saying. Fourteen days later, the tracking stations spotted another object being launched and then monitored it into a very low orbit. It was suspected that it was an attempt to launch another Space Station, but this was flatly denied by the Russians, who said that it was a 'test satellite', and it burned up in the atmosphere eleven days later.

With all this in mind, the Skylab project, which had been born during the Apollo missions, came to fruition. Skylab was exactly what the name suggested: a laboratory in the sky. It was built from the Saturn S-IVB stage of the Saturn V rocket booster left over from the Apollo programme. It consisted of a two-storey accommodation, 48ft long and 21ft in diameter. The upper section was the laboratory/workshop, whilst the lower section contained living quarters for three astronauts.

The workshop had two large solar panels attached to the outside which, during launch, were folded against the sides. Once in orbit the solar panels were extended. Attached to the forward end was the airlock module. This enabled the astronauts to enter and leave the workshop area without having to depressurise it. Also in the airlock module were the temperature controls, air purification systems, electrical controls and hazard warning systems for Skylab. Attached to the aft end of the workshop were the multiple docking modules that enabled the Apollo spacecraft to dock at either of two docking ports, one on the end and the other on one of the sides.

The living quarters consisted of three bedrooms, a kitchen and a bath; the crew had a veritable warehouse full of clothes consisting of 60 pairs of jackets, trousers and shorts; 15 pairs of boots and 210 pairs of underpants. The bathroom contained over 50 bars of soap, 50 towels and 1,800 urine and faecal bags. The latter were used to carry out investigations on the bodily wastes of the astronauts for medical and biological reasons. In fact, the size of Skylab was almost the same as that of a small two-bedroomed house.

The first stage would be launched complete with all necessary stores and equipment, and once in orbit would be visited by numerous astronauts. As it turned out, only three crews would ever visit Skylab before it was deemed to be redundant and sent into a decaying orbit around the Earth.

Right from the start there were problems. The lift-off of Skylab I from Cape Canaveral on 14 May 1973 went as normal, but within 1 minute of leaving the pad Flight Controllers of the Saturn V reported 'a strange lateral acceleration'. Within minutes an endless stream of telemetry confirmed there was an issue. A metal shield only 0.025in thick which fitted around the orbital workshop area had torn loose. This shield was a critical part of the workshop's protection, inasmuch as it was there to protect the spacecraft from small meteoroids and also the searing heat of the sun. If that wasn't bad enough, radar trackers in Australia reported that the two solar panels of the workshop had not deployed. This meant that there was no power going to the Space Station.

As the sun beat down on the unprotected spacecraft, temperatures within started to soar to 190°F. Controllers on the ground issued a command to Skylab to start rotating slowly so as to dissipate the heat, and engineers on Earth devised a kind of parasol that could be fitted over the top of the workshop area. When the first astronauts, Commander Charles Conrad Jr, Paul J. Weitz and Dr Joseph P. Kerwin, all from the US navy, arrived on board Skylab II, that was their first task. Donning their EVA suits, Conrad and Kerwin climbed outside Skylab, fitted the parasol and then released the solar panels, which immediately gave power to the station. Skylab was operational and proved conclusively that no matter how sophisticated technology became, when major problems arose it was left to man to fix them. With Skylab stabilised the first mission was to last twenty-eight days, in which a number of experiments were to be carried out. It was soon realised that the amount of experiments would have to be reduced because of the maintenance factor concerning Skylab itself. The repairs carried out initially would have to be continually monitored. The crew returned to Earth in their Apollo spacecraft having made the Skylab ready for the next crew.

The second crew, Alan L. Bean (Commander), Jack R. Lousma and Owen K. Garriott, spent fifty-six days orbiting the Earth aboard Skylab II. They carried out a variety of experiments and continued to maintain the station.

The third and final mission, Skylab III, with Gerald P. Carr (Commander), William R. Pogue and Edward G. Gibson aboard, carried out the final selection of tests and proved conclusively that man could live in space in relative comfort. Skylab was allowed to fall into a decaying orbit around the Earth and burnt up on re-entry some years later. When the idea of a space laboratory was first put forward, there were a number of sceptics who had doubts about the usefulness of such a vehicle. Among these was the director of Kitt Peak Observatory, Leo Goldberg, who said at the end of the last mission:

> Many of us had serious doubts about the scientific usefulness of men in space, especially in a mission such as the ATM (solar observatory), which was not

> designed to take advantage of man's capability to repair and maintain equipment in space. But these men performed near-miracles in transforming the mission from near-ruin to total perfection. By their rigorous preparation and training and enthusiastic devotion to the scientific goals of the mission, they have proven the value of men in space as true scientific partners in space science research.

This statement summed up to perfection the feelings of the American scientists and reinforced NASA's commitment to the space programme.

Meanwhile, in Russia, after a break of over two years, the Soyuz programme re-started with the launch of Soyuz 12 at 1218 hours GMT on 27 September 1973. The crew of Vasiliy Grigoyevich Lazarev and Oleg Grigoryevich Makarov were the first Russians to venture into space since the tragic deaths of the crew of Soyuz 11 two years previously. The primary mission of this spaceflight was to evaluate the newly designed spacecraft, and after two days in space they returned safely to Earth on 29 September at 1314 hours GMT, landing 440km south-west of Karaganda.

The success of this mission caused the next flight to be brought forward, and on 18 December 1973, just three months later, Soyuz 13 blasted off the launch pad at Baikonur and into orbit. The crew, cosmonauts Petr Ilich Klimuk and Valentin Vitalyevich Lebedev, carried on board their spacecraft the large Orion-2 astrophysical camera. They carried out astrophysical observations of stars in the ultraviolet range and spectrozonal photography of certain areas of the Earth. In addition to these experiments and investigations, they continued to test the systems aboard the new spacecraft. After nearly eight days in space, the spacecraft returned to Earth and landed in the middle of a snowstorm 200km south-west of Karaganda on 26 December 1973. The rescue team that found them were surprised to see them alive, so severe was the weather at the time.

The launch of Salyut 3 on 25 June 1974 placed another of the Russian Space Stations in orbit. The improved model and the experience obtained were placing the Russians well ahead of the Americans. Salyut 3 had three sets of solar panels grouped around the core instead of having two sets of solar panels that opened like wings fore and aft.

Soyuz 14 was the next Russian spacecraft to blast off the launch pad at Baikonur at 1851 hours GMT on 3 July 1974. The crew, cosmonauts Yuri Petrovich Artyukhin and Pavel Romanovich Popovich, carried out a successful docking with Salyut 3 during their fifteen-day mission. On board the Space Station they discovered a large number of improvements including an exercise treadmill, a small library of books and a wide variety of food and drink. The hierarchy began to realise that if cosmonauts were to spend lengthy periods in space their living and working conditions had to be made as comfortable as possible.

The two cosmonauts installed four cameras, two in the lower part and two in the upper part of the station, and a solar telescope where the operator stood at its

Crew of Soyuz 12: Vasiliy Grigoyevich Lazarev and Oleg Grigoryevich Makarov.

base and operated it by means of a control panel. The remainder of their mission was classified and no details are available.

The spacecraft returned to Earth on 19 July 1974, touching down within 2km of the estimated landing area which was 140km south-east of Dzkezkazgan.

One month later on 26 August 1974 at 1958 hours GMT, Soyuz 15 lifted off the launch pad with cosmonauts Lev Stepanovich Demin and Gennadi Vasilyevich Sarafanov aboard. Their primary mission was to rendezvous with Salyut 3 and carry out a series of investigations and experiments. According to official sources the joint experiment was a total success, but other more reliable sources said that the mission was a total failure. It appears that the automatic docking system failed and the mission had to be aborted, hence it only lasting two days. The spacecraft returned to Earth on 28 August.

The American and Russian crews of the joint ASTP.

The space race, although on the face of it appearing to create a chasm between the United States and Russia, did in effect draw the two countries closer. A joint American/Russian space mission started to take shape when on 2 December 1974 Soyuz 16, with cosmonauts Anatoly Valisyevich Filipchenko and Nikolay Nikolayevich Rukavishnikov aboard, lifted off from Baikonur and into space. Their mission was to evaluate the on-board systems of the Soyuz spacecraft, which had been made compatible with those of the American Apollo spacecraft for the forthcoming Apollo Soyuz Test Project (ASTP).

What had brought this mission about was the concern of both countries of the use of space, and it was decided that some form of co-operation was needed. Earlier President Kennedy had written to Russian Premier Nikita Khrushchev about the setting up of a world weather satellite system and the exchange of acquired data concerning the Earth and the space surrounding it. This brought some limited response from the Russians, but after the Apollo XIII incident it was decided that in the event of any spacecraft getting into difficulties there should be some common ground that would enable either country to launch a rescue mission.

On 26 December 1974 Salyut 4 was launched. The following day Soyuz 17 was launched with cosmonauts Lieutenant-Colonel Alexi Gubarev and civilian

Georgi Grechko aboard, and docked with the Space Station. The two cosmonauts carried out the usual checking of all the on-board systems, including the testing of a piece of equipment that allowed the cosmonauts to obtain drinking water from the cabin atmosphere by condensing it. A number of other experiments were carried out over the next twenty-nine days, and on 9 February at 1103 hours GMT the spacecraft landed back on Earth 110km north-east of Tselinograd. On 24 January 1975 the now defunct Space Station Salyut 3 re-entered the Earth's atmosphere over the Pacific and burnt up.

As if to tell the Americans that they were ahead again in the space race, the Russians launched Soyuz 18A at 1102 hours on 5 April 1975. The two cosmonauts, Vasiliy Grigoyevich Lazarev and Oleg Grigoryevich Makarov ran into problems soon after launch. Their primary mission was a rendezvous with the Salyut 4 Space Station, but soon after the first stage had separated the second stage of the rocket fired but failed to separate. With the second stage still firing, the crew demanded that the flight be aborted but the Ground Controllers could see nothing wrong in their telemetry signals. After much arguing, the second stage was separated by a Ground Control command at a height of 192km. An emergency re-entry was initiated, sustaining 20.6+G as it passed through the atmosphere and finally crashed into the Altai Mountains. The spacecraft then tumbled wildly down the mountainside coming to rest on the edge of precipice. The only thing that prevented it from hurtling over the side to almost certain death was the parachute, which snagged on a tree.

Initially the crew thought that they had landed in China, but they were discovered sitting beside the capsule by some local people – speaking Russian. Vasiliy Lazarev suffered internal injuries from the 20G re-entry and the tumble down the mountainside, and was never to fly again. All cosmonauts when they flew in space were given a spaceflight bonus of 3,000 roubles, but because they had not completed their flight, Lazarev and Makarov were at first refused payment. It took appeals right up to President Leonid Brezhnev to get the decision repealed.

The American ASTP patch.

A proposed joint mission between the two super-powers was suggested, with the link-up in space of an Apollo and a Soyuz spacecraft. After a great deal of difficulties, language and trust being just two, it was realised that both countries were going to have to be completely open about their space technology and the designs of their spacecraft.

The plans for the joint American/Russian ASTP mission were almost in place when Soyuz 18 was launched at 1458 hours GMT on 24 May 1975. The crew, Colonel Petr Ilich Klimuk and Vitaliy Ivanovich Sevastyanov docked with the Salyut 4 scientific orbital station and carried out a number of experiments and investigations during their sixty-three-day mission. As with all the Russian missions, information regarding the types of medical and biological experiments and investigations are extremely sketchy, and most of the time non-existent. It is known, however, that they did carry out in-depth investigations of the constellations Virgo, Scorpio and Cygnus using the X-ray telescope. The spacecraft returned to Earth on 26 July 1975 at 1418 hours GMT, landing 56km east of Arkalyk.

The results of the Soyuz 16 ASTP mission were the final piece of the space jigsaw that culminated on 15 July 1975 at 0820 hours EDT, when Soyuz 19 lifted off the launch pad at Baikonur and into space. On the other side of the world on the 15 July 1975 at 1550 hours EDT a Saturn 1B rocket, with Apollo XVIII on top, lifted off the launch pad at the Kennedy Space Center and into space. The Apollo/ Soyuz Test Project (ASTP) crews consisted of Soyuz cosmonauts, Colonel Aleksei

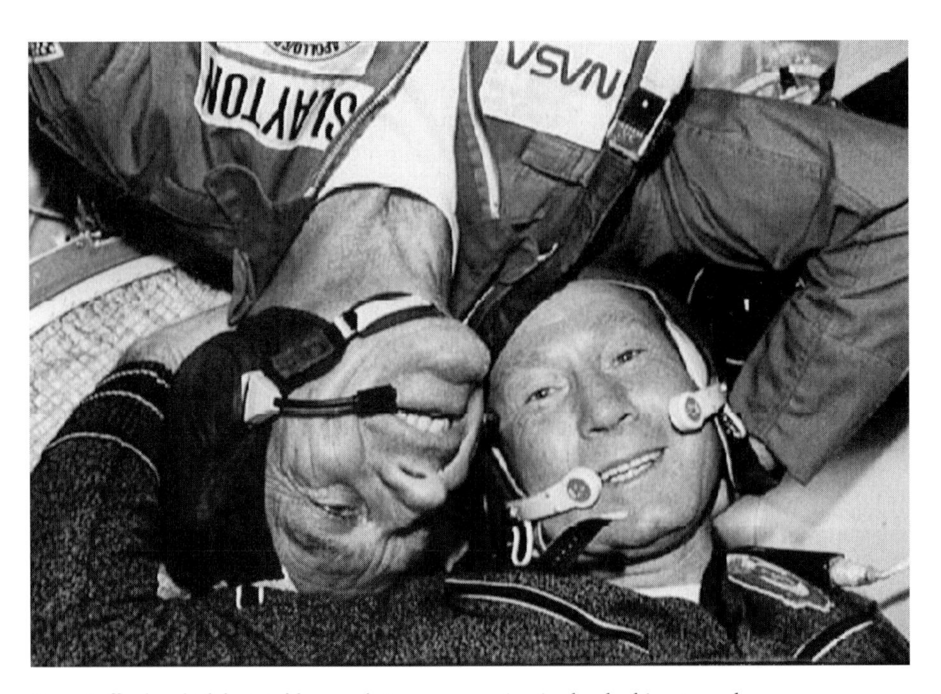

Tom Stafford and Aleksei Arkhipovich Leonov meeting in the docking tunnel.

Arkhipovich Leonov and civilian engineer Valeriy Nikolayevich Kubasov, and Apollo astronauts, Tom Stafford, Vance DeVoe Brand and Donald 'Deke' Slayton.

Such was the interest in this historic meeting that the launch of the Soyuz spacecraft was broadcast live around the world, something that had never happened before in the history of the Russian space programme. In fact, very few westerners have ever witnessed the launch of a Russian spacecraft or satellite. The two spacecraft, now in orbit, edged towards each other; then, at a height of 140 miles above the Atlantic Ocean, they met. Minutes later the two spacecraft docked together and history was made. On the television screens in Mission Control, Houston, Texas and in Soyuz control at Kaliningrad, Russia, there appeared the close-up of a hatch. Suddenly the hatch opened and the smiling face of a Russian cosmonaut Aleksei Arkhipovich Leonov appeared, then a hand stretched out and the face of an American astronaut Tom Stafford appeared on the screen. The two men shook hands warmly and greeted each other. The Apollo/Soyuz link-up cemented the friendship between the Russian and American astronaut/cosmonaut fraternity that had sprung up during training. Later both crews got together and carried out the first ever in-space press conference.

The mission lasted nearly six days, during which a number of experiments were conducted. One experiment carried out by the Russians involved three small cylinders containing metals that could not be uniformly mixed on the ground because of the Earth's gravity. On Earth the heavier of the metals, in their molten state, settled to the bottom before the mixture cooled and solidified. The astronauts took the experiment, loaded the cylinders into an electric furnace in the docking module and proceeded with the sequence of heating, melting and cooling the samples in the weightless conditions. The test samples were returned to the cosmonauts for their scientists to evaluate the results. There were a number of medical experiments also carried out covering mainly biological investigations.

There was almost a serious problem for the Apollo crew when, just before splashdown, the crew was gassed by nitrogen tetroxide that had escaped from the RCS (Reaction Control System) thrusters. Although Vance Brand was rendered unconscious and the remaining two astronauts became very woozy, they quickly recovered and none of them suffered any ill effects from the experience.

When the two spacecraft rendezvoused in space they opened up another chapter in man's desire to explore the universe. The whole mission was a complete success for both technology and the world. The American astronauts said afterwards that the construction of the Russian spacecraft looked more like a plumber's nightmare than a sophisticated spacecraft. But what was apparent was that there was a true *esprit de corps* between the astronauts and the cosmonauts. East and West could work harmoniously together in the furtherance of space exploration.

CHAPTER SIX

THE DEVELOPMENT OF THE SPACE SHUTTLE

After the success of the Mercury, Gemini, Apollo and Skylab missions, and the excitement of men walking on another world had died down, the space programme in America suddenly appeared to come to a standstill.

A great deal of thought had been given to the cost and the waste of materials that constituted the launch of the Mercury, Gemini and Apollo spacecraft. From the massive rocket that left the launch pad, all that returned to Earth was the tiny capsule that had been perched on top. The majority of the remaining parts of the rocket burnt up on re-entry, but some pieces became part of the massive junkyard that circled the Earth. However, NASA had not been idle: since the beginning of the 1960s they had been carrying out experiments with lifting bodies. These were small wingless aircraft equipped with a rocket motor, which were dropped from a converted Boeing B-52 bomber. The object of the exercise was to see if a re-usable spacecraft could be developed that could manoeuvre in space and then land back on Earth.

NASA had also been carrying out tests in 1963 using the North American X-15 rocket plane, taking the aircraft to heights and speeds never before experienced. At one time twelve test pilots, including Neil Armstrong, Forest Peterson and Scott Crossfield, carried out 199 flights, finally reaching altitudes of 98,800m (324,200ft) and speeds of 4,520mph (Mach 6.7). This took the aircraft virtually into space where it behaved much like a Mercury space capsule using small jets in the nose and, in the case of the X-15, in the wingtips, to control the pitch, roll and yaw. It was the use of small jets like these that was to later enable the Space Shuttle Orbiter to manoeuvre whilst in space.

The research aircraft were designed to withstand temperatures of 1,200°F, but on a number of occasions this was exceeded and temperatures of 2,000°F were recorded. These aerodynamic heating investigations also helped to create a new glass for the windshield. Originally the glass was made of a soda-lime mixture, but this failed on a test flight and a new windshield, made of alumino-silicate, was created. The X-15 provided invaluable information for pilots of very high-speed

Neil Armstrong beside the North American X-15 when working as a test pilot for NASA.

aircraft concerning the control surfaces and their reaction under high-speed stress and temperatures.

In Russia the Mikoyan-Guryevich Company (MiG) was working on the development of an aerospace system, known as the SPIRAL OS. It consisted of a re-usable hypersonic launch aircraft, an expendable two-stage rocket booster and an orbital spaceplane. The hypersonic launch aircraft was in the format of a large arrow-shaped flying wing, powered by four turbo-ramjet engines. On top of this was placed the orbital spaceplane and the expendable two-stage rocket. The theory was that the launch aircraft would take the spaceplane and the two-stage rocket to a height of 30km. The launch aircraft would then release them and return to its base. The two-stage rocket would then ignite and push the orbital plane into space. The rocket released from the spaceplane would be left to burn up as it re-entered the atmosphere.

After a number of wind tunnel tests and design alterations, it was decided to end the project because of the massive financial drain it was creating. The fact that it was under-funded from day one did not help, but the SPIRAL OS cosmonaut training programme was ended and the spacecraft was used to test analogue systems.

In April 1969 NASA had formed a committee to evaluate a proposal called Phase A, which was for a two-stage space transportation system that consisted of an Orbiter and a re-usable Booster craft. Both craft would have to have the capability of returning to Earth under their own power after launch. They would be launched vertically, powered by high-performance, oxygen-hydrogen throttleable engines. Once the correct altitude had been reached the two craft would separate, the Orbiter crew would then power up and continue into space, whilst the Booster crew would fly their craft back to base and land as a conventional aircraft.

With the idea of using a reusable spacecraft fixed firmly in their minds, NASA asked for proposals from the major aircraft companies. Grumman, McDonnell Douglas and Rockwell submitted designs and proposals and all were remarkably similar. On 26 July 1972, Rockwell was appointed prime contractor to build the Space Shuttle and on 9 August 1972 they were given the authority to proceed. The wings of the Orbiter were to be made by Grumman, whilst the tail was to be made by Fairchild. NASA continued with the lifting body programme, finishing the tests in 1975.

In Russia, an unmanned Soyuz spacecraft, Soyuz 20, was launched on 17 November 1975 with a number of biological experiments aboard. It also carried

North American X-15 rocket plane in flight.

Russian Orbiter Buran with escorting MiG-25 fighter after a test flight.

a series of tests in which the spacecraft performed some dockings and undockings in preparation for the use of the unmanned Progress supply spacecraft.

The Russian designer Vladimir Chelomei resumed work on his spaceplane although a design developed by NPO Energia called the Buran had been accepted as the prime spacecraft in the development of a re-usable Space Shuttle system. Chelomei's spaceplane, known as the LKS, was selected as a back up for the Buran, but was not completed until 1979. This included the construction of a full-scale mock-up of the LKS, which, according to some sources, was completed in less that a month and shown to the military hierarchy in an attempt to get the Buran project cancelled. The reverse happened; it was the Buran project that was accepted and in 1981 the development of the LKS was stopped and the whole project put on hold. There may have been some element of truth in the bitter arguments that Chelomei had been having with the Kremlin hierarchy, as in 1991 the workshops of NPO Mashinostroyeniye, where the spaceplane was being developed, were broken into and everything, including the mock-up, plans and machinery, was destroyed. It is thought that the KGB was behind the incident, but in Russia at the time anything that involved state security was surrounded by secrecy.

The design of the Buran Space Shuttle Orbiter was almost identical to that of the American Space Shuttle Orbiter. Initially Russian design engineers had dismissed the US design because of the large wings, and had carried out extensive aerodynamic tests whilst developing the Soyuz spacecraft. The results from these tests had shown that the large wings severely restricted the weight of the payloads that could be carried because of the weight penalties incurred by the wings themselves. They favoured the lifting body design with virtually no wings. After a great deal of debate and consideration of a large number of different designs, it was decided reluctantly that the American design was by far superior to anything that they could come up with. As one NPO Energia designer/engineer said:

> There is no point in picking a different inferior design just because it is original, that just defeats the object.

A new sub-contractor was chosen to build the spacecraft, the famous Russian aircraft manufacturer Mikoyan-Guryevich (MiG) who had been involved in the SPIRAL OS project. They created a new design bureau called Molinya to create the Buran spacecraft and carry out all the necessary tests.

The Russians also continued with their Salyut programme with the launch of Salyut 5 Space Station on 22 June 1976. The first of the Soyuz spacecraft to visit the station and dock was Soyuz 21, with cosmonauts Colonel Boris Valentinovich Volynov and Lieutenant-Colonel Vitaliy Mikhaylovich Zholobov on board. The launch, at 1209 hours GMT on 6 July 1976, went ahead with no difficulty and Soyuz 21 entered orbit at a height of 274km. On the second day in orbit the two spacecraft closed and docked. The mission was scheduled to last two months and a number of experiments, including the use of a hand-held spectrometer to study aerosol and industrial pollution in the Earth's atmosphere, had been carried out with the orbital research station, when Zholobov became unwell. The illness had started after they had been in space for about thirty days and had got progressively worse. After forty-nine days in space, with the majority of the experiments and investigations completed, it was decided to return the two cosmonauts to Earth. The spacecraft landed 200km south-west of Kokchetav at 1833 hours GMT on 25 August 1976. It was never officially disclosed what illness Zhobolov had suffered from, but a psychological breakdown was one theory and space sickness another. It is telling that he never flew again (space sickness usually lasts from one to three days from the start of the mission).

The following month the Russians decided to make use of one of their spare Soyuz ASTP spacecraft, and in place of the Apollo/Soyuz docking unit, they fitted a Zeiss camera. With cosmonauts Vladimir Viktorovich Aksyonov (a civilian engineer) and Colonel Valeri Fedorovich Bykovsky on board, Soyuz 22 lifted off the launch pad at Baikonur at 0948 hours GMT on 15 September 1976.

The mission of Soyuz 22 was to test and perfect a number of scientific-technical instruments that could be used for studying the geological characteristics of the Earth's surface from outer space. When perfected, it was also argued that similar methods could be used when exploring other worlds for economic purposes. With all the experiments and scientific investigations completed, the spacecraft returned to Earth at 0742 hours GMT on 23 September 1976 and landed 150km north-west of Tselinograd.

Crew of Soyuz 22, Vladimir Viktorovich Aksyonov and Colonel Valeri Fedorovich Bykovsky, carrying out the traditional signing of their spacecraft on landing.

In the United States the development of the Space Shuttle and its first reusable spacecraft, the Orbiter, was about to be unveiled. The Space Shuttle programme had been developed at the cost of a second Skylab, and the remaining pieces of Apollo hardware were sent to museums. The estimated cost of developing the Space Shuttle was $5.15 billion, although it actually overran to the tune of $6.744 billion, which was less than a quarter of the total cost of the Apollo space programme. Then problems arose regarding the weight of the Orbiter, which ended up nearly 20 per cent heavier than had been estimated. The result was that the US Air Force payloads, which had been one of the mainstays behind the programme, could not be carried into orbit. The USAF later developed its own programme using a re-designed Titan 4 expendable rocket to carry its payloads.

The shuttle programme also inherited fixed costs from the now-defunct Apollo project, in the shape of government and contractor workers at the Kennedy Space Center and at other sites. The cost to keep all the facilities and the people to run them in place was a staggering $2.8 billion per year and that was without a single flight leaving the space centre. Added to this was the cost of each flight of the Space Shuttle, which was around $100 million per year. This figure was significantly higher than that mooted by NASA when the programme was first introduced.

The Space Transportation System (STS) Space Shuttle consisted of a main external tank, 154.2ft in length and 27.5ft in diameter, to which two Solid Rocket Boosters (SRB) 149.16ft in length and 12.16ft in diameter were attached. Fixed to this was a glider aircraft called the Orbiter (OV), 122.2ft in length, a height of 56.67ft and with a wingspan of 78.06ft. The length of the cargo bay in the Orbiter was 60ft with a diameter of 15ft. Although the Orbiter had three main engines that gave out 393,800lb of thrust each on lift-off, on re-entry and landing it had no power. It was one of the largest gliders ever built, the exception being the Second World War German glider, the *Messerschmitt* Me 321B-1, with a wing span of 180ft 5in.

The first Space Shuttle Orbiter was named Enterprise and rolled out from Rockwell's assembly facility at Palmdale, California on 17 September 1976. It was transported by road, a distance of 36 miles, to the Dryden Flight Research Center at the Edwards Air Force Base, California for the Approach and Landing Tests (ALT). Two crews were assigned to carry out the tests, astronauts Fred Haise, Gordon Fullerton, Joe Engle and Richard Truly. The tests were conducted throughout 1977 and consisted of five 'captive' flights in which the Orbiter was mounted on top of a specially converted Boeing 747, known as the Shuttle Carrier Aircraft (SCA).

The Boeing 747 Shuttle Carrier Aircraft (SCA) was originally an American Airlines Boeing 747 (100 series) aircraft and, after it had accumulated 8,899 flying hours and carried out 2,895 landings, it was purchased by NASA for the princely sum of $15 million. It had been originally used at the Dryden Facility for the investigation of wake vortex problems experienced by wide-bodied aircraft. For the shuttle project the galleys and the majority of the passenger facilities were removed and additional instrumentation was installed in the cockpit to register the mid-air separations and additional weight whilst the 'piggy-back' rides were in progress. The crew also had a slide escape system installed, in the event of problems during mid-air releases. An unusual piece of equipment was installed inside the fuselage; this was a mass inertia damper, weighing 450kg, which moved laterally on rollers that were set into the floor of the fuselage. The idea of this damper was to counteract the oscillations that might be caused by the turbulence over the 747's tail after air had flown over the captive Orbiter. The four Rolls Royce RB211-524B2 engines were replaced with Pratt and Whitney JT9D-7AHW engines, increasing the thrust required.

The aircraft's main structure was reinforced to support the Orbiter's weight of 150,000lb and was fixed to the top of the aircraft by means of three supports, one forward and two aft. The same connections would be used to fix the Orbiter to the main fuel tank of the shuttle. The aircraft was flown by a senior NASA test pilot, Fitzhugh Fulton Jr, a former pilot of the Boeing B-52 that was used to launch the X-15. He was also the project pilot of the XB-70 bomber Valkyrie and the YF-12A Blackbird high-altitude reconnaissance aircraft. All the other members of the crew were experienced test pilots and test engineers.

The Russians, in the meantime, were developing their Space Station programme at considerable speed. With Soyuz 22 still in space, Soyuz 23 was already

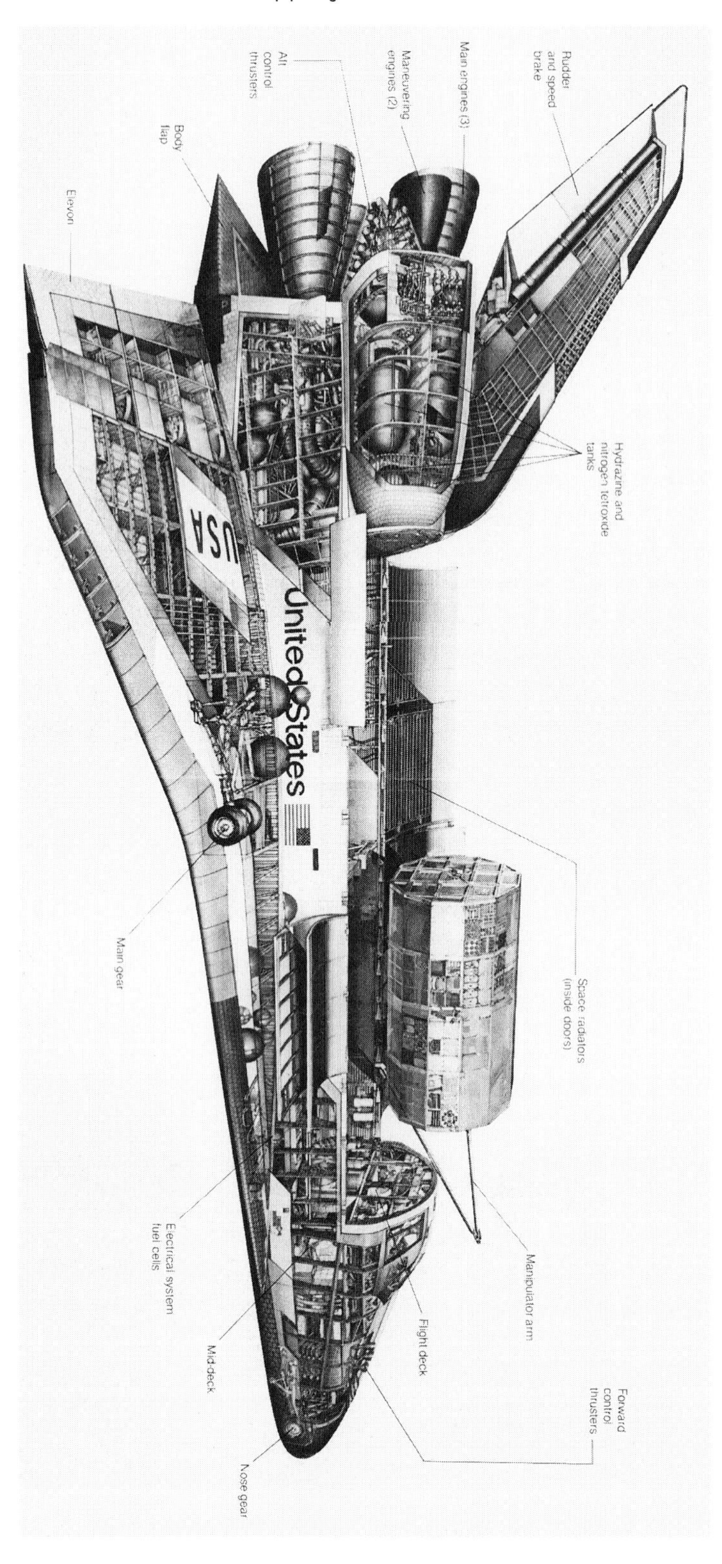

Cutaway drawing of the Orbiter spacecraft.

The Orbiter spacecraft Enterprise being taken along the road from Palmdale, California, to the Dryden Flight Test Centre – a distance of 36 miles.

The engineless Orbiter Enterprise on the back of the converted Boeing 747 SCA.

on the launch pad being prepared for another mission to Salyut 5. Its two cosmonauts, Lieutenant-Colonel Valeri Ilich Rozhdestvenskiy and Lieutenant-Colonel Vyacheslav Dimitriyevich Zudov, clambered into the cramped capsule to take them to what was to be their home for the next couple of months and at 1740 hours GMT on 14 October 1976 lifted off the launch pad at Baikonur and into orbit to rendezvous with Salyut 5. As the spacecraft attempted to dock with Salyut 5 a fault was discovered with the main antenna of the Igla rendezvous system. After repeated unsuccessful attempts to dock the two spacecraft the mission was aborted and the Soyuz spacecraft returned to Earth. But the drama didn't end there. The spacecraft landed in Lake Tengiz during a horrendous blizzard. Frogmen managed to attach a flotation collar to the capsule but the rescue boat was unable to reach the crew because of the partially frozen surface of the lake. Divers managed to get a line on the spacecraft and connect it to one of the helicopters, which then pulled it into the shore. After spending almost half the night trapped in their capsule, the two cosmonauts were released. When the rescuers reached them they were surprised to find the two men still alive.

The next Soyuz mission was on 7 February 1977 when Soyuz 24 docked with the Salyut 5 Space Station. At 1612 hours GMT on 7 February Soyuz 24, with Yuri Glazkov and Viktor Gorbatko aboard, lifted off the launch pad and into orbit around the Earth.

Their mission was to carry out a series of scientific experiments very similar to those of Soyuz 21. This was one of the most successful Soyuz missions, in terms of experiments and investigations accomplished, to that date. The flight was scheduled for seventeen days, and within that time the two cosmonauts accomplished as much as the crew of Soyuz 21 did in fifty days. At 0938 hours on 25 February 1977 the spacecraft landed just 37km north-east of Arkalyk.

In America the Approach and Landing Tests (ALT), which included the first 'captive' flights of the Space Shuttle Orbiter, were carried out in June and July 1977 without incident. The initial tests were concerned with taxiing trials with the Shuttle Carrier Aircraft (SCA) with the Orbiter attached, to determine the ground handling, braking and structural loading up to the point of take-off.

The Orbiter Enterprise was attached to the converted American Airlines Boeing 747-100 and a seventeen-piece tail cone was fitted over the aft end of the Orbiter's fuselage, covering the area where the rocket engines would be. This was to make the early flights as aerodynamic as possible. The first five flights were carried out with the Orbiter unmanned; this was followed by three further mated flights. There were two crew members aboard the Orbiter but none of the flight systems were powered up. The first flight, crewed by Fred Haise and Gordon Fullerton, remained attached to the 747 throughout and with the tail cone attached. The second flight crewed by Joseph Engle and Richard Truly was the same, as was the third flight crewed by Haise and Fullerton.

Originally six flights had been planned but so successful were the first manned flights that only three were carried out. Finally five manned 'free' flights were planned where the Orbiter was released, by means of explosive bolts, from the supports. Two of the flights were with the tail cone in place, followed by three with the tail cone removed. The Orbiter made all the flights without incident and glided down to landings at the Edwards Air Force Base with a landing speed between 213 and 226mph.

Whilst the Americans continued with the development of the Space Shuttle, the Russians kept up their investigations in the Salyut scientific research complex. The next Salyut Space Station was to see a milestone inasmuch as it was to be the first to use the two docking units on the complex simultaneously. This meant that when a crew went aboard to carry out experiments and investigations they could be joined by another team some months later who would then take over, thereby releasing the first crew so that they could return to Earth. Another team would replace the remaining crew aboard the station, some months later, each one overlapping its predecessor.

On 29 September 1977 Salyut 6 was placed into orbit around the Earth. It was made up of three cylinders connected by conical adapters. It consisted of a transfer module, a work compartment, an intermediate chamber, equipment bay and a scientific bay. The living space had an area of 100 cubic metres, whilst the transfer module was bounded by a sealed cylindrical shell just 2m in diameter and a sealed cone-shaped shell. The conical part was fitted over the top of the passive cone-shaped docking mechanism (the active part of the mechanism was on the transport spacecraft). A hatch was fitted into the conical structure, which allowed the crew to enter and leave the station for EVA purposes.

A large amount of equipment was fitted on the outside of the Space Station: handrails and anchoring points, by which cosmonauts in their spacesuits attached themselves to the station whilst they carried out servicing and repairs; panels for studying micrometeorite particles and the soiling of external optical surfaces; external television cameras; spherical tanks of the gas composition support system that contained a supply of compressed air; and sun and ion orientation sensors for the station attitude control system.

The transfer module, which housed the cosmonauts' spacesuits, attachments and control panels, was also the air-lock chamber. The module had seven port-holes, some of which were fitted with astro-orientation instruments, which, together with their responding controls and control panels, maintained the orientation of the Space Station.

The work compartment was made up of five posts. Post No 1 was the main control area for the primary control systems and was situated in the lower part of the compartment. There was just enough room for two cosmonauts to work side-by-side. The post also contained the regeneration cartridges of the gas composition support system and the cooling and drying units of the thermal control system.

Post No 2, also called the astropost, was the astro-navigation and astro-orientation facility section. Between the two posts was the area where the cosmonauts had their meals at a small table fitted with special appliances for heating food.

No 3 post was equipped to operate all the scientific apparatus and was located in the large diameter section of the work compartment close to the far end of the post. As with all the posts there were facilities for anchoring and communication. All the food supplies were kept in containers in the instrument section. At the far end of the upper section of the work compartment were the toilet and shower facilities, and close by were two air-lock chambers used for the removal of biological waste products. They were collected in special containers and ejected into space where they burnt up in the atmosphere.

Post No 4 was located in the lower central part of the work compartment, where the majority of the medical experiments were carried out and where the photographic and television equipment was housed. On both sides of No 4 post were the cooling and drying units of the thermal control system, the electronic units of the station's Attitude and Motion Control System (AMCS) and the radio equipment. The AMCS was one of the most vital systems on the Space Station because without it Salyut 6 would be continuously rolling and turning in an uncontrolled manner. The water regeneration and scientific apparatus control panels were housed in Post No 5.

There was a gap of some nine months before the next Soyuz flight. Soyuz 25, with Vladimir Kovalyonok and Valeri Ryumin aboard, lifted off at 0240 hours GMT on 9 October 1977. Their mission was to dock with Salyut 6 and carry out a series of experiments aboard the scientific research complex. After repeated unsuccessful attempts to dock with Salyut 6 the mission was aborted. The spacecraft returned to Earth at 0325 hours GMT on 11 October after only two days in space. It is not known what the problem was regarding the docking, but a similar one had befallen Soyuz 23.

The next mission, Soyuz 26 with crew members Yuri Romanenko and Georgi Grechko, was launched at 0119 hours GMT on 10 December 1977 from the space centre at Baikonur. Because of the problems Soyuz 25 had experienced trying to dock with the forward port of Salyut 6, the crew of Soyuz 26 docked aft. After docking, the two cosmonauts transferred into it to start their experiments and investigations and became the first cosmonauts to do so. During a spacewalk to inspect some suspected damage to the primary docking port thought to have been caused by Soyuz 25 when it docked, Yuri Romanenko, who was not tethered to the spacecraft, missed his hold and started to float away. Fortunately for him Georgi Grechko, who was tethered, stretched out as far as he could and managed to catch hold of Romanenko before he got too far from the spacecraft. Although he did not have his safety lines attached, Romanenko still had his communications umbilical cable attached, so the danger was not so dramatic as it at had first appeared.

One month later, on 10 January, the crew of Soyuz 27, Vladimir Dzhanibekov and Oleg Makarov, joined the crew of Soyuz 26. On 20 January 1978 the Space Station was the first to receive supplies of fuel, equipment, instruments and materials from an unmanned supply spacecraft: Progress. This was the first time that three spacecraft had been docked together with the Space Station.

The four crew members worked together to carry out a series of experiments before the Soyuz 26 crew returned to Earth using the Soyuz 27 spacecraft on 16 January 1978. On landing, Romanenko told Grechko that his father had died two weeks earlier, but it had been decided not to tell him until he was back on Earth because of the psychological distress it may have caused him. Grechko later agreed that it had been the right thing to do and was pleased that his friend Yuri Romanenko had insisted that he be the one to tell him.

A new type of EVA suit was designed for the cosmonauts. It had an all-metal body piece with flexible soft material for the arms and legs. The joints had sealed bearings and articulated joints that corresponded with the arm and leg joints of the wearer. The semi-rigid suit was the first of its kind to be used in a space environment. It had a number of advantages over the usual EVA suit inasmuch as it was much easier to get on and off and the hermetic sealing provided greater security.

Then on 2 March 1978 the Russians made space history when an international crew arrived at Salyut 6 aboard Soyuz 28. The two cosmonauts, Aleksei Gubarev and Vladimir Remek of Czechoslovakia, joined the two resident cosmonauts, Vladimir Dzhanibekov and Oleg Makarov, and carried out extensive tests on natural resources. They also completed a photographic survey of the central and southern parts of the USSR, as it then was. They made a large number of medical and biological research investigations. One of their investigations, called Extinkstia, involved the observation of changes in the brightness of the stars as they disappeared below the horizon. The information gathered from this research was required for a study of the dust layer that had been formed by micrometeorites at a height of 80–100km above the Earth. After seven days of intensive work, Gubarev and Remek returned to Earth on 10 March 1978. During their time on the station the cosmonauts were supplied with materials and necessities by Progress, the unmanned supply spacecraft.

America and Russia were not the only countries in the world looking toward the stars. In 1978 an article in the Chinese technical journal *Navigation Knowledge* announced that Zhen Xinmin, head of the Chinese Space Agency (up to this point no one even knew that China had a space agency), and his team was working on putting a man in space. Not only that, but they were also working on the development of a Space Station at the same time.

This immediately interested the Americans and they asked if a team of engineers and scientists from NASA could visit the Chinese Space Agency. This was arranged and everything was shown to the American team except the hardware.

'We can all theorize, but where is the reality?' was one comment passed by a NASA engineer after being told that what they had been shown was all there was. In was later discovered that, in the five years preceding 1978, China had launched five recoverable satellites, none of which was capable of carrying a man. But the signs were there that their intentions were serious. Soon after, Wang Zhuanshan, Secretary General of the New China Space Research Society and Chief Engineer of the Space Centre of the Chinese Academy of Sciences, declared that although they now had the technology to launch and recover a man from space the cost had proved prohibitive.

One year later the Chinese released photographs of astronauts training at the space centre along with drawings of a rocket fitted with an escape tower. An escape tower is only necessary if a flight is intended to carry human cargo. But what was becoming increasingly clear was that, since the death of Mao Zedong in 1976, in-fighting for the leadership had retarded scientific and technological advancement in virtually every field, including space exploration. It was not until 2003 that the Chinese managed finally to put a man into space.

CHAPTER SEVEN

THE SPACE SHUTTLE MAKES ITS APPEARANCE

In the United States in the late 1970s, the Space Shuttle programme was starting to take shape. In March 1978 the Orbiter Enterprise was ferried on top of the Boeing 747-100 SCA to NASA's Marshall Spaceflight Center in Huntsville, Alabama, to undergo a series of mated vertical ground vibration tests. These were completed by the end of the month and the Enterprise was ferried to the Kennedy Space Center. There it was mated to the external main fuel tank and the two solid rocket boosters (SRBs), and transported on the mobile launch platform to Launch Complex 39A. The Orbiter served as a practice launch facility to check the various aspects of the launch. In August the Enterprise was ferried back to NASA's Dryden facility in California atop the 747 SCA, and from there back to Rockwell's Palmdale Final Assembly Facility in October, this time by ignominious means on the back of a truck. The Enterprise was built purely as a test vehicle and was never equipped to go into space. A second SCA, a Boeing 747-100R, was acquired to relieve the workload of the first SCA.

One of the greatest challenges that faced NASA engineers was the development of the thermal protection for the Orbiter. The heatshield that had protected the Mercury, Gemini and Apollo spacecraft on re-entry was known as an ablative shield and could only be used once, whereas the Orbiter's heatshield had to be used time and time again. In addition to this was the weight factor: the ablative shield on the earlier spacecraft weighed 100lb per cubic foot, whilst the thermal tiles weighed a mere 9lb per cubic foot. The ablative heatshield burnt off very slowly, whereas the Orbiter's retained the heat whilst on re-entry then dissipated it almost instantly as it entered the cooling atmosphere of Earth. The leading edges of the wings and nose area were the parts most subjected to re-entry heat and were each constructed in a single piece. These were made of reinforced carbon-carbon and were grey in colour; they were able to absorb heat at temperatures up to 3,000°F. The remainder of the Orbiter's skin, looking like a patchwork quilt, was covered with a mixture of black and white tiles. The black tiles were able to absorb temperatures of between 1,200–2,300°F whilst the white tiles could absorb heat up to 1,200°F. Such were the properties of the black tiles and the speed in which they dissipate heat, that they could be heated until they were glowing red and then be safely picked up with the bare hand. Tests on these tiles were originally carried out by fitting them to the wings of an F-15 fighter and an F-104 Starfighter, both aircraft belonging to NASA.

The construction of both the black and white tiles consisted of mixing fibres of pure white silica, which was refined from common sand, with de-ionised water and a number of other chemicals. The mixture was then poured into a plastic mould and then subjected to great pressure to extract all the excess liquid. After drying the damp blocks in microwave ovens they were placed in a sinter oven. This oven heats the tiles to a temperature of 2,350°F, fusing the fibres without melting them. The tiles are cut roughly to shape before the final finish is carried out using diamond-tipped cutters and rotary profile grinders. The tiles were then spraycoated, glazed and waterproofed, creating a tile, which was amorphous borasillicate glass. This was the basic formula for both types of tiles, the only difference being that of the external coating of different chemicals which defined the area on the shuttle that was to be subjected to the greatest heat.

Over 70 per cent of the Orbiter's external surface was shielded by 30,761 individual black and white tiles – the black tiles being placed where the highest temperatures would be reached. When the tiles were fitted to the skin of the Orbiter small felt pads called 'strain isolator pads' were fitted between the tile and the skin. This was because the tiles, although able to sustain the varying temperatures easily, were very fragile, and during flight, when the Orbiter was subjected to both extreme temperatures and aerodynamic pressures, the metal of the spacecraft would twist, shrink and expand. The thermal protection for the Orbiter, unlike all the other systems that had been tested on the unmanned flights, could only be tested properly on a flight into space and back. The lives of two astronauts depended on the expertise of NASA's engineers and the trust they had in them. There were a few minor problems, however, when it was discovered on landing that a number of tiles were missing. This was soon rectified by using a different, stronger fixative. The Space Shuttle programme was under way.

Two launch sites had been selected, the Kennedy Space Center in Florida and the Western Test Range at the Vandenberg Air Force Base, California. The launches requiring equatorial orbital trajectories would be launched from the Kennedy Space Center, whilst those requiring polar orbital planes would be fired from the Western Test Range. The launch pads themselves (39A and 39B) were left over from the Apollo launch days and had to be modified to accommodate the Space Shuttle, as did the mobile launch platforms, also known as the 'crawlers', which transported the Space Shuttles from the VAB to the launch pad.

There are two crawler transporters at the Kennedy Space Center, both built by the Marion Power Shovel Company in Ohio in 1965 for the Apollo launches.

The Space Shuttle on the crawler platform moving towards the launch pad.

The crawlers are self-propelled vehicles that travel at 1mph when carrying a Space Shuttle and 2mph when empty, using one gallon of fuel every 20ft. Powered by two 2,750hp diesel electric engines, and consisting of a flat rectangular metal bed with four sets of caterpillar tracks similar to those used on tanks in each corner, the crawler weighs, when carrying the Space Shuttle, a staggering 11 million pounds. The crawler has a crew of twenty-six including two drivers, one at each end of the vehicle, and can move the tracks independently, enabling the vehicle to move forwards, backwards and sideways. At the end of 2004, the 'shoes', all 456 of them, each weighing 2,200lb, were replaced at a cost of $10 million; cracks were found in some and it was decided to renew them all.

The launching of the Space Shuttle from the backs of the crawlers was one of the most dangerous moments of the whole mission. What has to be remembered is that the Orbiter, when mated with the main external tank and the two Solid Rocket Boosters (SRB), is, for all intents and purposes, a giant flying bomb. Once the SRBs had been ignited, for the next 3 minutes of flight they could not be switched off and could not be discarded until empty. The main external fuel tank was made up of two tanks that contained 141,000 gallons of liquid oxygen in one and 385,000 gallons of liquid hydrogen (-423°F) in the other. When the two were combusted, the temperature reached 6,000°F, leaving the whole launch sequence fraught with danger. On the first flights, when there were only two crew members, ejector seats were fitted, but when the missions got under way with additional members of the crew aboard, these were removed.

The crawler platform on the launch pad roadway.

The sequence of events for the mission would start at the launch where the Space Shuttle, after lifting from the pad, would be subjected to maximum dynamic pressure 60 seconds later on reaching 33,600ft. Two minutes after launch, with the speed of the shuttle at 3,094mph and at an altitude of 28 miles, the two Solid Rocket Boosters (SRBs) would separate from the main external tank. At an altitude of 68 miles, at a speed of 17,440mph and 8 minutes after launch, the main engine would cut off and the external tank would separate from the Orbiter. The Orbiter would then continue into an orbit insertion, the altitude varying depending on the mission.

The missions then would swing into operation averaging between four and fourteen days, at altitudes of between 115 and 690 miles above the Earth. When the mission completed the Orbiter would decrease speed by means of the Orbital Manoeuvring System (OMS) and the Reaction Control System (RCS) thrusters in the nose and tail, and then enter the Earth's atmosphere. The landing would be at either the Kennedy Space Center or at the Edwards Air Force Base. Other emergency landing sites were available around the world. The Orbiter was scheduled for a two-week turnaround but, under certain circumstances where there was more than one Orbiter, it was hoped this could be measured in a matter of days. In fact it was never achieved. The fastest time that could safely be considered was ninety days.

The Russians, in the meantime, had spent the months preparing for another mission, but it wasn't until 15 June 1978 that Salyut 6 was visited again, when Soyuz 29, with cosmonauts Vladimir Kovalyonok and Alexander Ivanchenkov aboard, docked with the Space Station. After switching on the air regeneration and thermal regulating systems, the two cosmonauts then carried out the necessary housekeeping duties required to make the Space Station habitable. They also had to reactivate the water recycling system so that the water left aboard by the previous crew could be reprocessed. With all these tasks completed they settled down to conduct a series of scientific investigations and experiments. Two weeks later they were joined by two more cosmonauts; Pytor Klimuk and Major Miroslaw Hermaszewski of Poland (Pol) aboard Soyuz 30.

Their task for the eight-day mission was to carry out joint Russian/Polish investigations into the process of obtaining semi-conductor materials under weightless conditions, medical and biological research into how spaceflight affected human organism and to photograph the Earth's surface and oceans. With these experiments completed the spacecraft returned to Earth on 4 July 1978.

It was during these interchanges between crews that it was decided that the crew leaving the Space Station would return in the spacecraft that had brought their replacements. This also enabled the Russians to carry crew members from other communist countries for either political of cultural reasons.

The crew of Soyuz 31, Valeri Bykovsky and Lieutenant-Colonel Sigmund Jähn of the German Democratic Republic (GDR) replaced the crew of the Salyut 6 Space Station, Vladimir Kovalyonok and Alexander Ivanchenkov, on 27 August. On 3 September 1978 the crew of Soyuz 29 returned to Earth in the Soyuz 31 spacecraft to a well-earned rest, after spending over seventy-nine days in space.

The Soyuz 31 crew were to spend the next sixty-eight days in space, and carried out extensive investigations and experiments covering a variety of medical-biological and technological programmes. Visual and photographic studies were made of different areas of the surface of the world's oceans using a special multispectral unit made by the Carl Zeiss Jena optical works in the GDR. With all their work and investigations completed, Bykovsky and Jähn returned to Earth in the Soyuz 29 spacecraft on 2 November 1978. The Salyut 6 Space Station had spent nearly thirteen months in space and orbited the Earth over 6,300 times.

The Salyut Space Station remained unmanned until the 26 February 1979 when Soyuz 32, with cosmonauts Vladimir Lyakhov and Valeri Ryumin aboard, arrived and docked. For the next 110 days the crew carried out a series of investigations and experiments, coupled with some necessary repair work and the obligatory housekeeping duties on the station itself. With the work completed the crew left the Space Station and returned to Earth on 15 June 1979. Earlier that year visits by a Hungarian, a Cuban and a Bulgarian cosmonaut had been cancelled on the grounds of 'cautious safety'; a mysterious term, the meaning of which even the Russian cosmonauts were not privy to.

The crew of Soyuz 35: Leonid Popov and Valeri Ryumin.

Nikolay Rukavishnikov and Georgi Ivanov of Bulgaria, aboard Soyuz 33 formed the next crew scheduled to visit the Space Station but, after launching from Baikonur on 10 April 1979 at 1734 hours, it failed to dock with Salyut 6. This was the third time this had happened and no explanation was ever given, but it was revealed some years later that the main engines on the spacecraft had had a 'burn-out'. The speed at which the Soyuz spacecraft was approaching the Space Station gave the crew a strong indication that something was wrong. Rukavishnikov, as commander, made the decision to fire up the reserve engine, but if that failed they would stay in orbit for another six days and then their oxygen would run out. The crew were left with two choices: either use the engine, if it fired, and dock with the Space Station and hope that a rescue craft could be sent, or use the engine to head back to Earth. The crew chose the latter option and fired up the engine.

The engine had to fire for more than 90 seconds or they would become a satellite. It did, and for a period of 213 seconds. The spacecraft landed 180km off course. The cause of the main engines not firing was never known.

Soyuz 34 was an unmanned spacecraft sent to collect Vladimir Lyakhov and Valeri Ryumin from Salyut 6.

Two cosmonauts, Leonid Popov and Valeri Ryumin, arrived on the Salyut 6 Space Station on 10 April 1980 aboard Soyuz 35, and during the next fifty-five days carried out maintenance duties and a number of undisclosed experiments. This was Popov and Ryumin's second visit to the Space Station.

Soyuz 36 carried the fifth international cosmonaut, Bertalan Farkas of Hungary and the Russian commander Valeri Kubasov to the Space Station. For the next fifty-six days the two men carried out the normal maintenance duties together with a number of experiments. The on-board Soyuz 35 crew returned to Earth using the Soyuz 36 spacecraft.

A new type of spacecraft made its appearance in 1980: the Soyuz-T model. Soyuz T-1 was an unmanned version. The first of the new manned spacecraft, Soyuz T-2, was launched at 1419 hours GMT on 5 June 1980 and docked with the orbiting Space Station Salyut 6 the following day. The crew, Vladimir Aksyonov and Lieutenant-Colonel Yuri Malyshev, carried out the testing of the on-board systems when they docked with Salyut 6. Although the mission was very short, just under four days, it served its purpose in testing the new model and the spacecraft landed back on Earth at 1240 hours GMT on 9 June 1980.

One month later Soyuz 37 was launched, with cosmonauts Colonel Viktor Gorbatko and Lieutenant-Colonel Pham Tuan, North Vietnamese air force pilot, aboard. After spending seven days aboard Salyut 6 carrying out a number of experiments, the two cosmonauts returned to Earth in Soyuz 36, leaving their spacecraft as a return vehicle for Leonid Popov and Valeri Ryumin.

The movement of crews to the Salyut Space Station was becoming almost like a bus service. Two weeks after the Soyuz 37 crew arrived back in Russia Soyuz 38 blasted off the launch pad at Baikonur.

On board was the seventh foreign cosmonaut to visit the Space Station Salyut 6, Arnaldo Tamayo Mendez of Cuba, the first Latin American in space, who was accompanied by Colonel Yuri Romanenko. During their seven-day stay they carried out a number of bio-medical tests concerning how the brain, blood circulation and eye functions of cosmonauts behaved under weightless conditions. The two cosmonauts returned to Earth on 26 September 1980 after completing their experiments.

The two remaining cosmonauts, Valeri Ryumin and Leonid Popov, guided Progress 11, one of the unmanned supply spacecraft, to the Space Station so that fresh supplies could be unloaded. With all the supplies stored away, the two cosmonauts prepared to leave the Space Station after 184 days in space. On 11 October 1980 the two cosmonauts climbed aboard their spacecraft Soyuz 37 and headed back to Earth landing close to Dzhezkazgan. The empty Salyut 6 Space Station was later pushed into a higher orbit by the supply spacecraft Progress 11, which was then undocked, headed back to Earth and burnt up on re-entry.

On 27 November 1980 Soyuz T-3 lifted off the launch pad from Baikonur, with three Russian cosmonauts, Leonid Kizim, Oleg Makarov and Gennadi Strekalov on board. The following day the spacecraft docked with the Salyut 6 where they carried out repairs and service on the Space Station. They also worked on a series of biological experiments and scientific and technical investigations before returning to Earth on 10 December after just twelve days in space.

CHAPTER EIGHT

THE FIRST RE-USABLE SPACECRAFT

It was to be four months before the Salyut 6 Space Station was visited again. On 12 March 1981, Soyuz T-4 was launched from Baikonur with cosmonauts Vladimir Kovalenko and Viktor Savinykh aboard. They were to carry out a number of repairs and general maintenance together with the normal housekeeping duties before being joined ten days later by the crew of Soyuz 39. The crew of Soyuz 39, Vladimir Dzhanibekov and Jugderdemidyin Gurragcha of Mongolia, were launched at 1900 hours GMT on 22 March 1981 from the Baikonur Space Centre aboard Soyuz 39. Their mission was to help to carry out the vital servicing and repairs to the Space Station and to undertake a series of medical and scientific experiments over the next three months. This was part of the INTERCOSMOS programme designed to train cosmonauts from Russia's satellite countries. The two-man crew docked with the Salyut 6 Space Station on 23 March and, after carrying out a series of experiments and investigations, returned to Earth on 30 March 1981. This left the crew of the Soyuz T-4 to carry out further experiments aboard the Space Station.

The development of the Energia-Buran spaceplane continued in Russia but it was plagued with problems. Throughout the 1980s test after test of both the launch rockets and the spaceplane continued. The first launch of a model of the Energia-Buran was scheduled to take place on 6 July 1984, but because of an electrical failure the model failed to separate from the launch vehicle. Five months later the first taxiing trials of the spaceplane were completed. It wasn't until 15 November 1988 that the one and only successful launch of the Energia-Buran spaceplane took place. The unmanned spaceplane separated from its launch vehicle and returned safely to Yubileinly airfield at Baikonur. One year later the huge Antonov-225 heavy transport aircraft was converted to carry the Energia-Buran.

In the United States NASA was preparing to launch the world's first re-usable spacecraft. Columbia was the second of the Orbiters, and was named after the first US navy frigate to circumnavigate the globe in 1846. Columbia was designed and constructed to fly into space and return. It was taken by the same method as

The Buran spacecraft on the back of the giant transport Antonov-225.

Enterprise to the Kennedy Space Center, to be mated with its external tank and SRBs and prepared for its first flight into space. The first shuttle astronaut crew was Commander John W. Young, already a veteran of four spaceflights (Gemini III and X, Apollo X and XVI), and Robert L. Crippen, pilot, on his first spaceflight. The back-up crew for this first flight were prime crew for the second, Joseph H. Engle and Richard H. Truly. The following Space Shuttle flights are in chronological order and not in order of mission number. There were several reasons for this, the main one being that in the later stages there were three Orbiters available and sometimes, if one of the Orbiters was scheduled for launch and experienced a major problem, it would be replaced by the next designated mission.

STS-1 Columbia (OV-102) blasted off the pad at the Kennedy Space Center on 12 April 1981 at 0703 hours EST. The launch had been scheduled for 10 April, but because of a problem in the Orbiter's general-purpose computer system when the back-up flight software failed to synchronise with the primary avionics, it was put back until 12 April.

The world watched and held its breath as the first Space Shuttle, STS-1, was launched into space and heaved a sigh of relief when, after spending two days in space testing all the systems, it touched down back on Earth. The spacecraft Columbia landed back at the Edwards Air Force Base, California, on 14 April, the first re-usable spacecraft to leave the Earth and return.

The spacecraft was returned to the Kennedy Space Center on the back of the 747 SCA (Shuttle Carrier Aircraft). This was the method that was to be used to ferry the Orbiter back to its launch site base wherever it landed.

The major systems all tested successfully throughout the flight, although some damage to the thermal tiles was suffered due to shock waves being created by the Solid Rocket Boosters (SRB). In all sixteen tiles were lost and 148 damaged on the first mission, and to resolve the problem modifications were made to the water sound suppression system on the launch pad. During launch, waterspouts that are located in the flame trench and on the platform itself released over 300,000 gallons of water during the 20-second period of the launch.

The launch of STS-1 on the 12 April 1981 coincided with the twentieth anniversary of Yuri Gagarin's flight on the 12 April 1961.

In the meantime, the Russians continued their own space programme with the launch of Soyuz 40. Dumitru Prunariu of Romania, together with Russian cosmonaut Leonid Popov, were launched from the space centre at Baikonur on 5 May 1981. The two cosmonauts docked their spacecraft with Salyut 6 the following day and joined the crew of Soyuz T-4. They conducted a series of investigations and experiments, undocked and returned to Earth on 22 May 1981. The Soyuz T-4 crew returned to Earth on 10 June, leaving the Space Station empty.

Six months after the success of STS-1 the second of the Space Shuttles, STS-2, lifted off the launch pad at the Kennedy Space Center.

STS-2 Columbia (OV-102), with crew members Joseph Engle and Richard Truly aboard, blasted off at 1009 hours EST on 12 November 1981. Like STS-1, it too had been delayed, but this time by over a month. It had been scheduled for launch on 9 October, but during the loading of the forward Reaction Control System (RCS) tanks, a spillage of nitrogen tetroxide occurred. The launch was then re-scheduled for 4 November, but during the final countdown a hold was called when low readings on the fuel cell oxygen tank pressures were discovered. Whilst these were being investigated it was found that in two of the three auxiliary power units (APUs) which operate the hydraulic systems, there were very high pressure readings. The APUs had to be flushed out and the filters replaced. Subsequently the launch was re-scheduled for 12 November.

STS-1 patch.

The crew of STS-1: John Young and Robert Crippen.

STS-1 launching – the first flight of the Space Shuttle.

STS-2 on the launch pad.

The Space Shuttle launched 2 hours and 40 minutes behind schedule on 12 November because a multiplexer/demultiplexer had to be replaced. During the launch it was noticed that one of the 'O' ring seals on the Solid Rocket Boosters suffered a 'blowby'. A 'blowby' is when fuel from the SRBs leaks out through the 'O' ring seal under immense pressure and is in danger of igniting. Tests were carried out on subsequent SRB 'O' rings at three times the pressure and found to be safe. (In 1986, however, when the STS-51B Challenger spacecraft blew up seconds after launch, it was this 'blowby' of the 'O' ring seal that was found to have been the cause.)

After entering orbit the crew of STS-2 settled down to carry out a number of tasks, including operating the remote manipulator system for the first time. On a Spacelab payload in the loading bay of the Orbiter was an experiment called Office of Space and Terrestrial Applications (OSTA-1), which when activated sent back a great deal of information to the mission scientists back on Earth. The five-day mission was cut to two days after the failure of one of the three fuel cells that produced electricity and drinking water.

The Orbiter landed at Edwards Air Force Base on 14 November 1981 and on subsequent inspection of the thermal tiles it was discovered that none had been lost, and only about a dozen had been damaged.

STS-2 was the last of the Space Shuttles to have the main fuel tank painted white; all future flights left it unpainted, giving the tank its now familiar orange colour and saving 600lb in weight.

Four months later the third shuttle flight was launched. The STS-3 Columbia (OV-102) mission was the longest of the shuttle flights to date. Columbia was launched into space at 1100 hours EST on 22 March 1982. The crew of Jack Lousma (Commander) and Gordon Fullerton (Pilot) were to carry out the most intensive mission yet, consisting of the continued testing of the Orbiter's systems, additional testing of the remote manipulator system and the measurement of thermal response of the Orbiter when placed in a variety of different attitudes to the sun. An OSS-1 pallet was installed in the payload bay to obtain information on contamination (gases, dust etc.) that may have been introduced into space by the Orbiter itself. A number of other successful scientific experiments were carried out during the mission, but there were a number of problems that affected the astronauts' comfort. These included issues with the thermostat that controlled the temperature within the spacecraft, a toilet that kept malfunctioning, space sickness and unexplained static that continually interfered with the astronauts' sleep.

Just prior to re-entry, the weather over the Edwards Air Force Base area deteriorated, so the landing site was changed to White Sands Missile Range, one of the designated alternative landing sites. Because of this an extra day was needed to change the re-entry position. When the spacecraft did eventually touch down at White Sands at 0904 hours on 30 March, after eight days in space, it encountered high winds which made the descent extremely difficult. It has to be remembered that the Orbiter was a glider and could only make one attempt at landing. After touchdown it was discovered that there was considerable damage to the brakes of the spacecraft and, due to a dust storm, extensive contamination damage to the tiles. This was the first and last time that the Space Shuttle landed at White Sands.

On the other side of the world, the continuing flights to the Russian Space Station were now being completely overshadowed by the flights of the Space Shuttle. These Russian flights were nevertheless important and the information they gathered, which was later shared with the Americans, would help pave the way for the construction of an International Space Station.

The launch of the Salyut 7 Space Station on 19 April 1982 demonstrated Soviet commitment to their own space programme. Although of the same design as the previous stations, Salyut 7 was plagued with technical problems throughout its lifespan. Two sections of the discontinued Almaz Space Station programme were mated with Salyut 7, which increased its size and allowed it to host a number of military experiments originally destined for Almaz. Salyut 7 was later replaced by the more successful Mir Space Station.

On 13 May 1982 at 0958 hours GMT, Soyuz T-5 with cosmonauts Anatoli Berezovoi and Valentin Lebedev on board blasted off the launch pad at Baikonur. This time it was to rendezvous with Salyut 7. After carrying out housekeeping duties they proceeded to conduct a series of medical and scientific experiments.

They were to spend almost 212 days in space before returning to Earth in the Soyuz T-7 spacecraft.

The crew of Soyuz T-6, Vladimir Dzhanibekov and Jean-Loup Chrétien of France joined the Soyuz T-5 crew aboard Salyut 7 on 24 June. The Soviet/French crew arrived to carry out a series of scientific and medical experiments, but only stayed for eight days before returning to Earth.

In the United States the R & D (Research and Development) shuttle flights were nearing the end of their flight tests; the last to be launched was STS-4.

STS-4 Columbia (OV102) was the final research and development mission, which also included one commercial experimental package, and blasted off Pad A at the Kennedy Space Center at 1100 hours EDT on 27 June 1982. This was the first of the shuttle flights to go ahead right on time and was a perfect launch. However, when the two SRBs separated from the external tank their parachutes failed to open and they broke up on impact with the water. It was also discovered that some rainwater had penetrated the protective coating of some of the thermal tiles whilst it was on the pad. On reaching orbit the crew, Thomas Mattingly (Commander) and Henry Hartsfield (Pilot), turned the affected area toward the sun to use its heat to dry it out. This was a serious issue because had they ignored the problem and the affected area been turned away from the sun, then the water would have frozen and expanded, damaging the tiles.

STS-4 crew: Henry (Hank) Hartsfield and Thomas Mattingly.

A number of experiments were scheduled for this mission, including a Department of Defence (DoD) payload. For obvious reasons details of these experiments are highly classified. Amongst the other experiments was the first commercial one that involved tests on a Continuous Flow Electrophoresis System (CFES), an Induced Environment Contamination Monitor (IECM), a Monodisperse Latex Reactor (MLR) and two Shuttle Student Involvement Programme (SSIP) experiments. The crew took numerous photographs of the Earth and of adverse weather systems. They also deployed a business systems satellite successfully, which emphasised the cost savings in using the Orbiter to launch various communications satellites.

After 113 orbits of the Earth the Orbiter re-entered the atmosphere and landed at Edwards Air Force Base, where President Ronald Reagan and his wife Nancy greeted the astronauts. The stage was now set for the Space Shuttle to become a viable and commercial enterprise.

Meanwhile, the Soviets were developing their Space Station programme and their crews were spending longer and longer in space. On 19 August 1982 Soyuz T-7 lifted off the launch pad and joined up with the Salyut 7 Space Station. The all-Russian crew of Leonid Popov, Alexander Serebrov and Svetlana Savitskaya were to spend the next 113 days in orbit conducting experiments and investigations. Svetlana Savitskaya was the second Russian woman to go into space, the first to carry out a spacewalk and the first to live on a Space Station. The Progress unmanned spacecraft delivered supplies and materials to the Space Station during their stay and again the spacecraft burnt up on re-entry after un-docking.

After joining up with the Space Station and the crew of Soyuz T-5, a number of experiments were carried out before the Soyuz T-5 crew returned to Earth on 1 September 1982. The Soyuz T-7 crew returned to Earth on 10 December 1982, leaving the Space Station empty.

The first of the commercial shuttle flights to deliver communication satellites commenced with the launch of STS-5 Columbia (OV102). This flight, the first operational mission, carried the largest crew up to that time – four astronauts. The fifth launch of the Orbiter Columbia took place at 7.19 a.m. EST, on 11 November 1982, and was the second on-schedule launch.

This was the first operational flight into space and carried two satellites, ANIK C-3 for TELESAT Canada and SBS-C for Satellite Business Systems. Each satellite was equipped with a Payload Assist Module-D (PAM-D) solid rocket motor. These motors fired 45 minutes after each payload had been deployed, placing the satellite into the pre-arranged elliptical orbit.

A planned spacewalk, the first for the shuttle programme, by Lenoir and Allen was postponed by one day after Lenoir became ill, but then had to be cancelled when the two spacesuits that were to be used developed problems. During the

STS-6 crew patch.

five-day flight, in addition to the launching of the two satellites, three SSIP experiments were carried out.

Columbia landed at Edwards Air Force Base on 16 November 1982. STS-5 was the first shuttle flight in which the crew did not wear pressure suits for the launch, re-entry and landing portions of the flight, similar in nature to Soviet Voskhod and Soyuz flights prior to the ill-fated Soyuz 11 mission. The mission was a success and the shuttle had proved that it was a viable concept and had tremendous commercial and military possibilities.

The crew of STS-5 was Vance D. Brand – Commander, Robert F. Overmyer – Pilot, Joseph P. Allen – Mission Specialist (MS) and William B. Lenoir – MS.

A new Orbiter spacecraft, Challenger (OV-99), appeared on 4 April 1983 when STS-6 lifted off from the Kennedy Space Center. It was the first use of a new lightweight external tank and lightweight Solid Rocket Boosters (SRB) casings. Originally scheduled for launch on 30 January, a hydrogen leak in one of the main engines was discovered. Then, after a flight readiness firing of the main engines on 25 January, fuel line cracks were found in the other two engines. A spare engine replaced the one with the hydrogen leak and the other two were removed, repaired and reinstalled.

The launch was further delayed when a severe storm caused contamination of the primary cargo for the mission, the first Tracking and Data Relay Satellite (TDRS). This occurred while it was in the Payload Changeout Room on the Rotating Service Structure at the launch pad and meant the satellite had to be taken back to its checkout facility to be cleaned and rechecked. The Payload Changeout Room and the payload bay also had to be cleaned.

Shortly after the TDRS had been deployed, its two-stage booster rocket tumbled out of control, placing the satellite into a low elliptical orbit. Fortunately, the satellite contained extra propellant beyond what was needed for its attitude control system thrusters, and during the next several months the thrusters were fired at carefully planned intervals, gradually moving TDRS-l into its geosynchronous operating orbit and thus saving the $100 million satellite. Peterson and Musgrave

Lightning striking the STS-8 launch pad.

carried out a 4-hour EVA to test the EVA suits, before returning to Earth. The crew consisted of Paul Weitz – Commander (A); Karol Bobko – Pilot (A); Donald Peterson (A) MS; and Story Musgrave (A) MS.

The launch of Soyuz T-8 on 20 April 1983 was the first failure to dock at a Space Station since Soyuz 33 in 1979. When the launch shroud separated from the booster, it took with it the rendezvous antenna boom. The crew, unaware that this had happened, believed the boom remained attached to the spacecraft's orbital module, and that it had not locked into place. Accordingly, they shook the spacecraft using its attitude thrusters in an effort to rock it forward so it could lock. The abortive docking attempts consumed much propellant. The crew shut down the attitude control system in order to preserve propellant. They then put the spacecraft into a spin-stabilised mode of the type used by Soyuz ferries in the early 1970s. With all control returned, the crew returned to Earth on 22 April, landing as normal. The crew was Vladimir Titov, Gennadi Strekalov and Aleksandr Serebrov.

The first American woman to fly in space was on 18 June when Sally Ride, a Mission Specialist, was aboard STS-7 Challenger (OV-99). This flight set the new record for the largest crew with five people on board: Robert Crippen – Commander; Frederick Hauck – Pilot; Norman Thagard – MS; John Fabian – MS; and Sally Ride – MS. Two communication satellites, Anik C-2 for Telesat of Canada and Palpa B-1 for Indonesia, were successfully launched during the first two days of the mission. Also carried was the first Shuttle Pallet Satellite (SPAS-l) built by Messerschmitt-Bolkow-Blohm of Germany. SPAS-l was unique in that it was designed to operate in the payload bay or could be deployed as a free-flying satellite. It was deployed by the Remote Manipulator System (RMS) and flew alongside and over Challenger for several hours while a camera took pictures from the SPAS-1 of the Orbiter performing various manoeuvres. The RMS later grasped the pallet and returned it to the payload bay.

STS-7 was scheduled to make the first shuttle landing at the Kennedy Space Center's Shuttle Landing Facility. However, unacceptable weather forced a change to the Edwards Air Force Base. The mission lasted 6 days, 2 hours.

On 27 June 1983 the launch of Soyuz T-9 was to be of some concern because of the booster failure that affected Soyuz T-8. Fortunately the launch went well and the separation of the capsule from the rocket was a smooth transition. Immediately after docking at Salyut 7's aft port, the crew of Vladimir Lyakhov and Aleksandr Alexsandrov commenced transferring the 3.5 tons of cargo lining its walls to Salyut 7. There was one moment of concern during the mission, and that was when a small object struck a Salyut 7 viewport. It caused a 4mm crack but did not penetrate the outer of the window's two panes. The Russians thought it was a fragment from a meteor shower, although it could have been a small piece of the orbital debris that was drifting around in space. It was the latter that was beginning to become an increasing cause for some concern, as each mission appeared to leave some sort of debris to drift into orbit.

Attached to the Soyuz T-9 was a Merkur Capsule and the crew loaded it with 350kg of completed experiments and unwanted hardware. The capsule was then undocked and returned to Earth.

STS-8 Challenger (OV-99) was the eighth shuttle mission, the third flight of the spacecraft Challenger and the first to carry an African-American, Guion Bluford. The flights were becoming more and more reliable as STS-8 blasted off Pad A at the Kennedy Space Center at 0232 hours EDT on 30 August 1983, only 17 minutes late due to bad weather when a lightning storm surrounded the launch area. After reaching orbit, the crew settled down to releasing the INSAT-1B, a multipurpose satellite for India. The satellite was attached to a Payload Assist Module-D (PAM-D) rocket motor and placed in its correct orbit.

The nose of the Orbiter was held away from the sun for a period of 14 hours to test the flight deck area in the bay in the extreme cold of space. The crew then filmed the performance of an experimental heat pipe operating under these conditions, also mounted in the cargo bay.

Higher loads were then placed on the manipulator arm to test the joint reactions to the extra weight. One of the experiments was to drop the Orbiter to an altitude of 139 miles to perform tests on thin atomic oxygen, in an effort to identify the cause of the glow that surrounded parts of the spacecraft at night.

Testing was conducted between the Tracking and Data Relay Satellite (TDRS) that had been launched by STS-6 and the Orbiter, using the Ku-band antenna. A number of other minor experiments were carried out just prior to the spacecraft making preparations for re-entry. STS-8 landed back at the Edwards Air Force Base at 0040 on 5 September 1983, the first time the Orbiter had landed at night. The crew consisted of Richard H. Truly – Commander; Daniel C. Brandenstein – Pilot; Dale A. Gardner – MS; Guion S. Bluford Jr – MS; and William E. Thornton – MS.

Three months later, on 26 September, as Soyuz 10A prepared to lift off, fuel spilled around the base of the launch vehicle and caught fire. The escape system was activated but such was the intensity of the fire that the control cables had already burned through. The crew, Vladimir Titov and Gennady Strekalov, unable to activate or control the escape system, frantically sought another way to get out, but 20 seconds later Ground Control was able to activate the escape system by radio command. Explosive bolts fired to separate the descent module from the Service Module and the upper launch shroud from the lower. Then the escape system motor fired, dragging the orbital module and descent module free of the booster at 14–17Gs of acceleration, which lasted 5 seconds. Seconds after the escape system activated, the booster exploded, destroying the launch complex. The descent module separated from the orbital module at an altitude of 650m, and dropped free of the shroud. It discarded its heatshield, exposing the solid-fueled landing rockets, and deployed a fast-opening emergency parachute. A safe landing occurred about 4km from the launch pad.

STS-9 Columbia (OV-102) was one of the most important flights of the Space Shuttle to date, as it not only carried six astronauts: John Young – Commander; Brewster H. Shaw – Pilot; Owen K. Garriott – MS; Robert Parker – MS; Byron K. Lichtenberg – PS; and the German Ulf Merbold – PS, the first astronaut from the European Space Agency on the first NASA/ESA sponsored Spacelab mission. The launch was scheduled for 30 September but whilst on the pad it was discovered that there was a suspect exhaust nozzle on the right SRB. The shuttle was returned to the Vehicle Assembly Building (VAB) and the Orbiter was de-mated from the external tank. The suspect SRB nozzle was replaced and the whole shuttle re-assembled. The countdown was started for 28 November.

The launch was perfect and the spacecraft Orbiter entered into orbit around the Earth. The primary interest of the mission was the investigation and capability of the orbital laboratory. The Spacelab was an orbital laboratory and observation platform and consisted of cylindrical pressurised modules and U-shaped unpressurised pallets, the latter of which stayed in the cargo bay during the flight.

A problem arose when Nos 1 and 2 general-purpose computers failed and No 1 inertial measurement unit also failed. It took over 8 hours to analyse and override the fault. More problems arose on landing when two of the three Auxiliary Power Units (APU) caught fire, but the spacecraft landed safely. Despite these problems the mission was deemed to be a success.

STS-41B Challenger (OV-99) was the next flight after STS-9; STS-10 was cancelled because of delays in getting the payloads ready. The next mission should have been designated STS-11, but it was decided to introduce a new numbering system and it became STS-41B. Vance Brand, the commander of the mission, was making his second spaceflight; his first had been on the groundbreaking Apollo/Soyuz flight.

After reaching orbit, the crew of Vance Brand – Commander; Robert Gibson – Pilot; Bruce McCandless II – MS; Ronald McNair – MS; and Robert Stewart – MS settled down to launch the two satellites, WESTAR-VI and PALAPA-B2. Problems started to arise when the Payload Assist Module-D (PAM-D), that was to put the satellites into their respective orbits, malfunctioned just after launch. This placed both satellites in extremely low Earth orbits. Things didn't get any better when the German-built Shuttle Pallet Satellite (SPAS) remained in the cargo bay when the Remote Manipulator System (RMS) suffered an electrical problem. The SPAS had originally been flown on STS-7 but had been recovered, refurbished and was due to be launched again.

The highlight of the mission, however, was when Bruce McCandless and Robert Stewart, using the Manned Manoeuvring Units (MMU), performed the first untethered spacewalks. They used the opportunity to practice procedures for retrieving the Solar Maximum Mission satellite which was scheduled to be retrieved and repaired on the next mission. During the mission Bruce McCandless carried out a spacewalk that left him almost alone in space.

Bruce McCandless on his untethered spacewalk.

Challenger landed back at the Kennedy Space Center, the first shuttle to land at the same site from which it had launched.

On 8 February the next mission to the Salyut Space Station, Soyuz T-10, blasted off from Baikonur. The Space Station was inert and the three cosmonauts, Oleg Atkov, Leonid D. Kizim and Vladimir Solovyov, entered it using torches; for the next sixty-three days they carried out the necessary housekeeping duties and repaired the propulsion system fuel lines. The repairs to the fuel lines meant that a number of EVA missions had to be carried out and it took almost a month to complete. They also installed a new solar array, which increased the electrical supply to the station.

The Russians continued to carry out missions to their Space Station and the transferring of crews. The launch of Soyuz T-11 on 3 April 1984 was the first flight of a cosmonaut from India, Rakesh Sharma. The other members of the crew were Yuri Malyshev and Gennady Strekalov. Sharma was to spend seven days on the Space Station conducting life science experiments before returning to Earth aboard the Soyuz 10B spacecraft.

Three days after the Russian Soyuz T-11 had lifted off, STS-41C Challenger (OV-99) launched from the Kennedy Space Center. This was the first direct ascent trajectory for the shuttle which reached its 533km-high orbit using the Orbital Manoeuvring System (OMS) engines only once to place it in its orbit.

The mission had two primary objectives. The first was to deploy the Long Duration Exposure Satellite (LDEF); the second was to capture, repair and redeploy the malfunctioning Solar Maximum Mission satellite, known as Solar Max, which had been launched in 1980.

On the second day of the flight, the LDEF was grappled by the RMS arm and successfully released into orbit. On the third day of the mission, Challenger's orbit was raised to about 300 nautical miles (556km), and manoeuvred to within 200ft of Solar Max. Astronauts Nelson and van Hoften, wearing EVA suits, entered the payload bay. Nelson, using the Manned Manoeuvring Unit (MMU), flew out to the satellite and attempted to grasp it with a special capture tool but failed. It began tumbling on multiple axes when Nelson attempted to grab Solar Max by hand, so the effort was called off.

The next day, Crippen manoeuvred Challenger back to Solar Max and Hart was able to grapple it with the RMS. They placed Solar Max on a special cradle in the payload bay using the RMS. They then began the repair operation, replacing the satellite's attitude control mechanism and the main electronics system of the coronagraph instrument. The ultimately successful repair effort took two separate spacewalks. Solar Max was deployed back into orbit the next day, thus concluding one of the most unusual rescue and repair missions in the history of the space programme. The crew consisted of Robert Crippen – Commander; Francis Scobee – Pilot; James van Hoften – MS; Terry Hart – MS; and George Nelson – MS. Challenger landed at Edwards Air Force Base after completing 108 orbits of the Earth.

Long Duration Exposure Facility (LDEF) being lifted from the payload bay of the Orbiter.

The Russians launched their second woman into space on 17 July 1984 aboard Soyuz T-12. Svetlana Savitskaya, Vladimir Dzhanibekov and Igor Volk carried out numerous experiments during their eleven-day mission, including the first EVA by a woman. Igor Volk was scheduled to be the pilot of the Buran spacecraft when it went into production.

A new spacecraft made its appearance in 1984 – Discovery (OV-103). It was to carry the crew of STS-41-D into space.

This was the first flight of the new spacecraft named after the ship *Discovery* in which Captain Scott and his crew went to the Antarctic in 1901. But again the mission was plagued with problems. Scheduled to be launched on 25 June, it had reached T-9 minutes when it was put on hold because of the failure of the Orbiter's back-up General Purpose Computer (GPC). The computer was replaced and the launch scheduled for the 26 June. At T-4 seconds the launch was aborted when the GPC detected a problem in the Orbiter's No 3 main engine. The Orbiter was de-mated from the shuttle and returned to the Orbiter Processing Facility (OPF) for the engine to be replaced.

The delay had caused other problems so STS-41F was cancelled and the payload re-assigned to STS-41D. The Orbiter was re-mated with the shuttle and returned to Pad A to be prepared for launch. Still troubles plagued the launch when, on the third attempt on 29 August, a discrepancy was noted in the master events controller of Discovery's computer software. This was quickly repaired and the launch scheduled for a fourth time, for 30 August. This time launch was delayed for 6 minutes because a private aircraft had strayed into the restricted area off the Cape.

At 0841 hours EDT on 30 August, STS-41D blasted off the pad at the Kennedy Space Center and after settling into orbit prepared to deploy three satellites. The satellites were the Satellite Business System SBS-D, SYNCOM IV-2 (aka LEASAT 2) and TELSTAR. Extending from the payload bay was the 102ft tall and 13ft wide Office of Application and Space Technology (OAST-1) solar wing carrying several different types of solar cells. It was designed to demonstrate and test a variety of lightweight solar arrays that could be used in the building and powering of future building in space, i.e. Space Stations.

Amongst the crew of Henry Hartsfield – Commander; Michael Coats – Pilot; Judith Resnik – MS; Richard Mullane – MS; and Steven Hawley – MS was Charles Walker. Payload Specialist Walker was the first fare-paying astronaut ever to fly in space. His company, McDonnell Douglas, paid NASA $60,000 to carry Walker aboard STS-41D so that he could carry out experiments on electrophoresis.

One problem occurred during the flight when the wastewater dump nozzle, a new design, froze up creating a large icicle. Commander Henry Hartsfield used the RMS to break it off just prior to re-entry. There was a danger, if it had

OAST-1 solar array about to be launched.

been left, that it would break off as the spacecraft re-entered the atmosphere and damage one of the insulating tiles that protected it.

With all the experiments completed, and after six days and ninety- seven orbits, the Orbiter prepared to re-enter the Earth's atmosphere. The landing site at Edwards Air Force Base had been chosen because it was the Orbiter Discovery's first flight and the runway there was ideal for a long runout. The spacecraft touched down at 0637 hours PDT on 5 September 1984.

Two months later, STS-41G Challenger (OV-99) lifted off the pad at the Kennedy Space Center. This was the first flight to have two women on the crew, Sally Ride and Kathy Sullivan. Sullivan was to carry out one of the EVAs, making her the first American woman to walk in space. The crew consisted of Robert Crippen – Commander; Jon McBride – Pilot; David Leestma – MS; Sally Ride – MS; Kathryn Sullivan – MS; Paul Scully-Power – PS; and Marc Garneau, a Canadian payload specialist.

Shortly after reaching orbit, the Earth Radiation Budget Satellite (ERBS) was deployed. This was followed by three experiments belonging to the Office of Space and Terrestrial Applications-3 (OSTA-3) from the payload bay. There was an experiment concerning the Orbital Refuelling System (ORS) demonstrating the possibilities of refuelling satellites in orbit.

A number of other experiments included the continuing RME and IMAX packages, a large-format camera, a package of Canadian experiments and eight Get Away Specials. All the experiments and EVAs were carried out successfully, and after eight and a half days and 133 orbits the Orbiter Challenger re-entered the Earth's atmosphere and landed at the Kennedy Space Center. There were a number of Department of Defence flights after this and information on these missions was never made available.

On the other side of the world, the Russians were preparing for their next space mission. The next visit to the inert Salyut 7 Space Station took place on 6 June 1985 when Soyuz T-13 was launched from Baikonur. The two cosmonauts, Vladimir Dzhanibekov and Viktor Savinykh, carried out a flyby inspection of the Space Station before finally docking. The solar array outside the station was pointing randomly as it rolled about on its axis and upon entering the station itself, they found that there was no electrical power and the walls and equipment were covered in frost. Eight of the solar-powered batteries were dead, the remaining two were useless, and the cosmonauts' first concern was to get the power back on. To do this they used their spacecraft to rotate the station until the solar arrays were pointing at the sun. Once this was done, the batteries slowly started to recharge. They then connected the Soyuz T-13's air generation system to the station's air heaters and slowly but surely brought the Space Station back to life.

The cause of the problem was discovered to have been the failure of a sensor in the solar array, and a problem in the telemetry radio system had not alerted the TsUP control. Once the frost had been cleared, the station's wall heaters were turned on to prevent the water from entering the equipment and causing further problems.

This really was a tremendous achievement as the cosmonauts had to work in freezing temperatures. It was not until the end of July that the station was regarded as being fully operational again.

The ninth expedition to the Salyut 7 Space Station caused a serious problem for the crew. Launched on 17 September, Soyuz T-14, with crew members Vladimir Vasyutin, Georgi Grechko and Alexander Volkov on board, docked two days later and joined the crew of Soyuz T-13. As the crews got down to the maintenance and housekeeping duties and started to carry out their experiments, Vasyutin began to feel unwell. At first it was thought he was suffering from space sickness, which usually clears up after a few days, but his health became progressively worse. By the end of October he was unable to help with any of the work aboard the Space Station, and none of the experiments he had been assigned had been done.

It soon became clear that Vasyutin needed medical attention and had to be returned to Earth as soon as possible. Scrambled communications with TsUP control confirmed this and the additional Soyuz spacecraft, which was attached to the Space Station, was utilised and on 14 November, Vladimir Vasyutin, Alexander Volkov and Viktor Savinykh returned to Earth. This left Georgi Grechko and Vladimir Dzhanibekov on Salyut 7 and they returned to Earth the following day. It was said that Vasyutin was suffering from a severe infection at the time.

With the success of the first commercial shuttle flight completed, NASA settled down to a programme of shuttle flights that would take the world of technology into a different era. These consisted of STS-51A, STS-51C, STS-51D, STS-51B, STS-51G, STS-51F, STS-51I, STS-51J, STS-61A, STS-61B and STS-61C. A number of communication satellites were deployed and many were 'captured' and repaired whilst in space. Among these flights were a couple of DoD missions, information on which was never released. The shuttle programme settled down with an air of confidence and a touch of complacency, but then came the tragic flight of STS-51L.

STS-51L – Challenger (OV-99) was to be unique inasmuch as it was to be the first time a private citizen was to be launched into space. Christa McAuliffe, a schoolteacher, had been selected after extensive and exhaustive tests and screening, to carry out the first teaching programme in space. The flight of STS-51L would also see the first loss of life in the American space programme since Apollo I.

The mission was beset with problems from the outset. It was scheduled for launch on 25 January 1986, but bad weather at the transoceanic abort landing (TAL) site in Dakar meant the alternative landing facility in Casablanca had to be used, but that was not equipped for a night landing. The launch time was then moved to a morning lift-off, but that was then aborted because the crews who processed the shuttle could not meet the deadline. A new launch date was scheduled for 27 January, but the gremlins struck again when a ground servicing equipment hatch closing fixture could not be removed. After a great deal of struggling the closing fixture was sawn off and the retaining bolt drilled out before the hatch could be closed.

The following day, 28 January 1986 at 0938 hours EST, the shuttle was delayed once again when the software that monitored the fire detection system in the launch processing computer failed during the loading of the liquid hydrogen.

Two hours later the shuttle lifted off and exploded in mid-air 73 seconds later, killing all seven members of the crew: Francis R. Scobee – Commander; Michael J. Smith – Pilot; Judith A. Resnik – MS; Ellison S. Onizuka – MS; Ronald E. McNair – MS; Gregory B. Jarvis –PS; and Sharon Christa McAuliffe – Teacher in Space Project.

The first indications as to how the accident had occurred came when photographic evidence was produced showing a strong stream of grey smoke spurting from the O-ring aft field joint on the right Solid Rocket Booster (SRB), at the moment the Space Shuttle lifted off the pad. This area of the SRB faces the external tank which houses the majority of the shuttle's fuel. As the shuttle moved upwards eight more puffs of grey smoke, each more black than the last, were seen coming from the area. The colour of the smoke and its density indicated that the grease, the joint insulation and the rubber O-rings in the joint seal were being burned by the hot propellant gases.

As the shuttle continued on upwards, 59 seconds after the launch, the black smoke disappeared only to be replaced by a plume of flame. Then the first indication that there was something wrong was picked up by Mission Control. Telemetry signals showed a difference in pressure between the right-hand SRB and the left-hand one. The right-hand SRB's pressure was considerably lower and 5 seconds later there was a distinct colour change in the plume of flame that was now streaming downwards, deflected by the aerodynamic forces. A rubber seal between two of the mated sections of the SRB had, because of the freezing

STS-51-L crew patch.

STS-51-L disintegrating.

weather, failed to seat properly and became stiff and unpliable. This led to hot combustion gases leaking out and igniting, resulting in flames like a welder's torch. At 72 seconds after launch the flame burnt through the SRB attachments which caused the SRB to break up. The main tank then ruptured and 3 seconds later the Space Shuttle exploded in a ball of flame, as the hydrogen in the external tank ignited.

It is an accepted fact that in any form of exploration there are going to be accidents, whether they are avoidable or not, but when it happens the shock is no less painful. As America tried to recover from the tragedy, the Russians launched Soyuz T-15 on 13 March 1986 to the empty Mir Space Station to collect some experiments that had been left on board. The crew of Leonid Kizim and Vladimir Solovyov were the first to return to the empty Space Station since Soyuz T-14. The crew then transferred all the experiments and their personal kit to Salyut 7. For the next fifty-one days the two men moved between the Space Stations, before returning to Earth on 16 July.

It was becoming clear that the existing Soyuz spacecraft was too small for the amount of material being transported to the Mir Space Station. A new Soyuz spacecraft, Soyuz TM, was designed and on 21 May 1986 an unmanned Soyuz TM-1 was launched.

With the success of the new spacecraft, the next mission to the Mir Space Station, Soyuz TM-2, was launched on 5 February 1987. The crew of Yuri Romanenko and Aleksandr Laveykin boarded Mir the following day. Romanenko was to spend 326 days aboard Mir, but Laveykin, who was scheduled to stay for the same period, developed heart irregularities which made it necessary to return him to Earth earlier than expected with Soyuz TM-3.

Soyuz TM-1 on its way to the launch pad at Baikonur.

Soon after settling in, the Kvant 1 module was launched to automatically dock with Mir. When the 11-ton module began homing in on Mir's aft port, the crew retreated to Soyuz TM-2 spacecraft so that they could escape in the event that the module got out of control. About 200m out the Igla homing system lost its lock on Mir's aft port Igla antenna. The cosmonauts watched from within Soyuz-TM 2 as the Kvant/FSM combination passed within 10m of the station.

The Kvant module drifted 400km from Mir before being guided back for a second docking attempt. Soft-dock occurred early on 9 April. Kvant's probe unit would not retract fully, preventing hard-docking between Mir and Kvant. The Soviets left Kvant soft-docked while they considered a solution. Manoeuvres were impossible during this period because the probe of the Kvant would wobble loosely in Mir's aft port drogue unit, banging the docking collars together.

On 11 April Romanenko and Laveykin carried out an EVA to examine and, if possible, repair the problem. They discovered a rubbish bag lodged in the docking unit, probably from the Progress 28 supply craft. On command from the TsUP, Kvant extended its probe unit, permitting the cosmonauts to pull the object free and discard it into space. Kvant then successfully completed docking at a command from the ground. The EVA lasted 3 hours and 40 minutes. The Kvant FSM undocked from Kvant on April 12, freeing the module's aft port to fill in for the Mir aft port.

The third expedition to Mir, Soyuz TM-3, was launched on 22 July 1987. On board Aleksandrov was to be Laveikin's replacement aboard Mir, becoming

Romanenko's new partner. Syrian guest cosmonaut Mohammed Faris and Soviet cosmonaut Aleksandr Viktorenko returned to Earth in Soyuz-TM 2 with Aleksandr Laveikin.

It wasn't until 21 December that the next flight to Mir took place when Soyuz TM-4 launched with a fresh crew for the Space Station. During the next few months they carried out the normal housekeeping duties required to keep the station running, including unloading Progress supply craft that kept Mir supplied with everything the crew needed. The returning crew almost always took the previous crews' spacecraft to return to Earth.

Soyuz TM-5 was the next spacecraft to visit Mir with a fresh crew on 7 June 1988, arriving on Mir the following day. Amongst the crew of Anatoly Soloviyov and Viktor Savinykh was the Bulgarian cosmonaut Aleksandr Aleksandrov, the second Bulgarian to be taken into space. The first Bulgarian, Georgi Ivanov, had been on Soyuz 33, but due to a technical problem the flight had been unable to dock with the Space Station.

With the normal housekeeping duties completed, work settled down on the astronomical experiments assigned to Aleksandrov.

Soyuz TM-6 took another non-Russian to the Space Station: cosmonaut Abdul Ahad Mohmand of Afghanistan. This crew was unique inasmuch as its commander, Vladimir Lyakhov, had been trained to fly the TM spacecraft solo in the event that a rescue spacecraft was needed to recover two cosmonauts from Mir. The third member of the crew was Dr Valeri Polyakov, who would remain aboard Mir with Titov and Manarov to monitor their health during the final months of their planned year-long stay.

The intention was that Mohmand and Lyakhov would return to Earth in the Soyuz TM-5 spacecraft, but during the descent Soyuz TM-5 suffered a combined computer software and sensor problem, which caused the re-entry rockets to stop their burn. After a quick checkup of position and attitude, Lyakhov realised there was nothing wrong and restarted the retroburn programme. Again the rockets failed to finish their burn and Lyakhov decided to wait for instructions from the ground. While waiting for a response, Abdul Mohmand discovered that even though the rockets failed to fire, the re-entry programme had continued and within a few moments the re-entry module would be separated from the orbital and engine modules, leaving them crippled in space. He warned Lyakhov and with less than a minute away from separation, Lyakhov stopped the programme. Re-entry occurred as normal on 7 September. After this the Soviets retained the orbital module until after de-orbit burn, as they had done on the Soyuz ferry flights.

Then, in August 1988, there was another disturbing incident, this time aboard the Russian Space Station during the Soyuz TM-6 mission, when an oxygen generator caught fire. The fire was smothered by Valery Polyakov using a spare uniform, but once again the matter was hushed up by the Russian authorities and only emerged when it appeared in a Russian scientific journal a couple of years later. It was then admitted, after a number of questions had been asked, that a similar fire had occurred aboard Salyut 6 in 1978. That fire had been serious and had almost caused the evacuation of the Space Station.

The first of the shuttle flights since the tragic accident that befell STS-51L lifted off from the Kennedy Space Center on 29 September 1988. STS-26 Discovery (OV-103) was on a mission to deploy a TDRS satellite amongst other experiments. There were two minor problems during the flight. One of these was the Flash Evaporator System, which was used for cooling the Orbiter, iced up and shut down. This increased the crew cabin temperature to approximately 87°F. It was not until day four of the flight that the problem was resolved. The second was when a Ku-band antenna for communications successfully deployed on day two, failed to respond properly and had to be stowed for the remainder of the mission.

One of the other experiments was the use of the Voice Control Unit (VCU). Discovery was the first spacecraft to fly in space equipped with a VCU, a 'computer' that recognised and responded to human speech. This speech recognition system controlled the cameras and monitors that were used by the crew to oversee the mechanical arm mounted in the cargo bay. Because of the experimental nature of speech recognition, this system was not used for any critical operations. There were initial problems which almost caused the tests to be cancelled, when it was discovered that the voice templates that were created prior to lift-off were found to have less than 60 per cent recognition for one crew member and less than 40 per cent recognition for another. The problem was resolved by retraining the templates. Further tests proved the VCU to be operational with a recognition success rate of over 96 per cent. The results concluded that weightless conditions caused a fundamental change in human speech which made making the templates prior to lift-off almost useless.

Two months later, on 2 December, STS-27 Atlantis (OV-104) lifted off on a DoD flight. As with all DoD flights no information is available. The crew consisted of Robert Gibson – Commander; Guy Gardner – Pilot; Richard Mullane – MS; Jerry Ross – MS; and William Shepherd – MS.

In Russia ground tests continued with the Buran Orbiter spacecraft, but interest in the project was waning and in 1992 funding for the programme was withdrawn. The only flight version of the Energia-Buran was destroyed when the roof of its hangar collapsed in 2002.

The Soyuz flights to Mir continued with regularity, taking a variety of Intercosmos cosmonauts on to the Space Station. Problems with keeping ageing Mir operational were beginning to concern both the Russians and the Americans, so in 1993, at a conference in Moscow, NASA negotiators came to an agreement with their Russian counterparts on the building of an International Space Station (ISS). Initially the American Space Station of Alpha and Russian Mir 2 would be merged, to be joined later by modules built by other countries.

STS-27 crew patch.

On 3 October 1994 the next mission to Mir was launched. Soyuz TM-20, with Alexander Viktorenko, Yelena Kondakova and Ulf Merbold aboard, docked with the Space Station's Kvant module. During the automatic docking procedure, the spacecraft started to yaw unexpectedly causing Viktorenko to take over manual control to dock the spacecraft.

In Russia, American astronauts were in training with Russian cosmonauts for forthcoming missions to Mir. One of these astronauts, Norman Thagard, was to be the first American to board the Russian Space Station. Together with Vladimir Dezhurov and Gennady Strekalov, he was to be launched aboard Soyuz TM-21, becoming a member of the Mir 18 crew.

During the training in Russia, Norman Thagard's relationship with his back-up crew member Bonnie Dunbar deteriorated. Thagard could speak and read Russian, but Dunbar struggled to the point that he had to translate almost everything for her. Because of this Thagard started to fall behind in his training so he refused to continue translating, obliging Dunbar to hire a translator. Back in Houston, this wasn't going unnoticed and rumours circulated that there were moves afoot to replace Thagard with Dunbar, but after a series of medical examinations back in Houston, where Bonnie Dunbar suffered an allergic reaction to a test, she was taken off the Mir programme by the Russians.

Soyuz TM-21 was launched from Baikonur on 14 March 1995 and docked with the Mir Space Station two days later. They docked at the Kvant module after the unmanned Progress M-26 had undocked and been placed in de-orbit to burn up on re-entry. The arrival of the three-person crew of Vladimir Dezhurov, Gennady Strekalov and Norman Thagard meant that there were thirteen astronauts in space at the same time, three already aboard Mir and seven aboard the Space Shuttle STS-67. On 22 March the Soyuz TM-20 spacecraft, with Alexander Viktorenko, Yelena Kondakova and Valeri Polyakov aboard, undocked from Mir and headed back to Earth leaving the crew of TM-21 in charge of the Space Station.

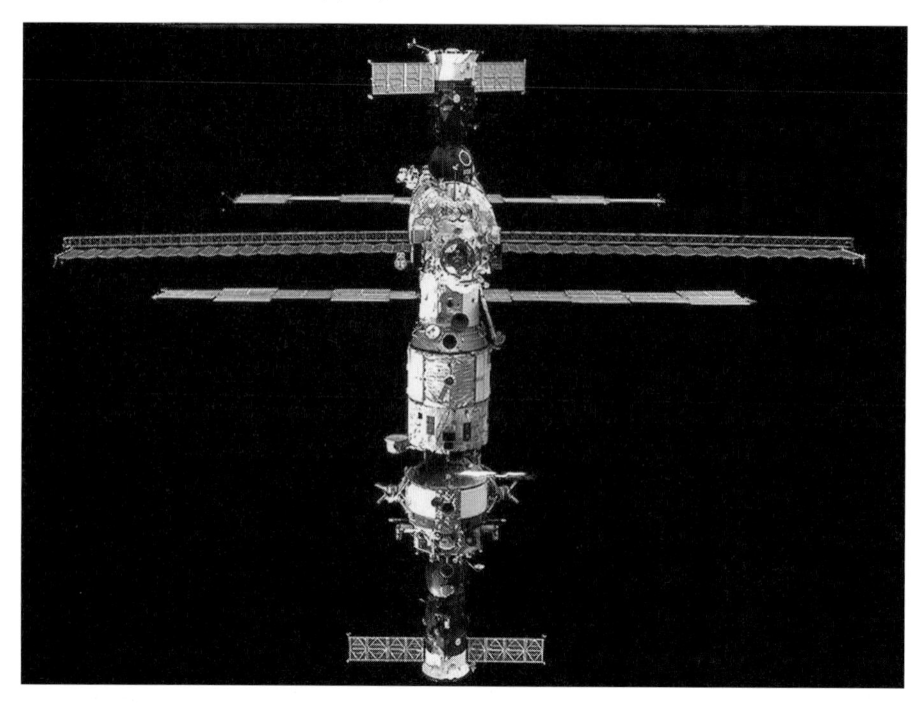

Russian Space Station Mir in orbit around Earth.

In America the shuttle programme continued to flourish and there followed a succession of shuttle flights: STS-29, 30, 28, 34, 33, 32, 36, 31, 41, 38, 35, 37, 39, 40, 43, 48, 44, 42, 45, 49, 50, 46, 47, 52, 53, 54, 56, 55, 57, 51, 58, 61, 60, 62, 64, 68, 66, 63 and 67. The flights were not always in numerical order for a variety of reasons. These missions deployed a variety of satellites and included the first docking of an American spacecraft with a Russian Space Station. This happened on 27 June 1995 when STS-71 Atlantis (OV-104) successfully docked with the Mir Space Station. This was the shuttle's third mission to Mir, only this time it docked with the Space Station. The mission was to deliver a relief crew of two cosmonauts, Anatoly Solovyev and Nikolai Budarin, to the station, and to recover American astronaut Norman Thagard.

The docking of the two space vehicles made this the largest spacecraft ever placed into orbit, the first ever on-orbit exchange of shuttle crew members and the one hundredth manned space launch by the United States.

There followed a further eleven missions, STS-70, 69, 73, 74, 72, 75, 76, 77, 78, 79 and 80, three of which, STS-74, 76 and 79, were to the Russian Space Station Mir.

STS-81 Atlantis (OV-104) was the fifth shuttle mission to Mir and the second one involving an exchange of US astronauts. Astronaut John Blaha, who had been on Mir since 19 September 1996, was replaced by astronaut Jerry Linenger.

Linenger spent more than four months on Mir. He returned to Earth on Space Shuttle mission STS-84.

In addition to the new Mir crew member, Atlantis carried the SpaceHab double module providing additional mid-deck locker space for secondary experiments. During the five days of docked operations with Mir, the crews transferred water and supplies from one spacecraft to the other. STS-81 would involve the transfer of 5,975lb (2,710kg) of logistics to and from the Mir, the largest transfer of items to that date. With the crew exchange completed, Atlantis undocked from Mir and returned to Earth.

The Russians continued to have problems aboard Mir. One in particular occurred in February 1997 when Soyuz TM-25 (with crew members Vasili Tsibliyev, Aleksandr Lazutkin and Reinhold Ewald aboard) collided with the station during docking, puncturing one of the modules.

Then, on 4 March 1997, an incident occurred which nearly destroyed Mir and its occupants. The unmanned Progress M-33 supply spacecraft was expected and the crew aboard Mir made preparations to receive it. What caused the problem has never been established, but the crew could not find the approaching spacecraft as it raced toward them until the very last minute and it passed beneath with only metres to spare. So concerned were the crew, that Linenger had been ordered by Vasili Tsibliyev to make ready the Salyut spacecraft as a 'lifeboat', and he was actually in the spacecraft when the unmanned Progress sped by. It wasn't until much later that NASA, after communication with Linenger, learned of the incident, which was indicative of the continuing secretive attitude still being adopted by the Russian hierarchy.

Linenger seemed to be more concerned in getting on with his experiments, and leaving the maintenance and repairs to the Russians, rather than working as part of the team. This wasn't helped by the mixed messages that were coming from the Mission Control in Russia who were monitoring the mission. There had been concerns voiced about sending Linenger to the Space Station because he was a loner who had his own agenda, and would only join in when it suited him. The Cold War between Russia and the United States had not completely thawed! The atmosphere on board the Space Station at the time was not conducive to good working relations.

The incident with the 'near miss' was followed by a dangerous on-board fire which took some time to extinguish and caused increased tension between the crew members. Alexander Lazutkin was installing one of the oxygen generating lithium perchlorate cartridges in the Kvant module. As he slotted the cartridge into place he heard a gentle hiss as the oxygen was released, then saw sparks appearing from the top of the cartridge. Suddenly the sparks ignited and plumes of flame spewed out in all directions. The fire was intense and lasted 15 minutes, shooting flames a metre long. A fire in a space environment was something that had never been experienced before, although both Jerry Linenger and Shannon

Lucid had conducted controlled experiments on a small scale whilst on shuttle flights. They had noted that the flames turned into a ball of fire, but Lazutkin only saw flames, fuelled by the oxygen, shooting in all directions.

The incident happened in an area between the crew and the two Soyuz spacecraft that were attached, preventing the crew getting to them had the fire got out of hand. During the fire, the smoke was so dense that everything had to be done by touch. On the ground this would have been difficult enough, but in space with no gravity, and everyone and everything floating around, it was ten times more difficult. Valeri Korzun ordered Aleksandr Lazutkin to ready one of the Soyuz spacecraft in case they needed to evacuate the station. In the meantime the remaining members of the crew grabbed all of the extinguishers they could find and passed them to Korzun, who continued to fight the fire. Some of the gas masks that were available were found to be faulty; something that the crew intended to 'mention' on their return to Earth during debriefing. Such was the intensity of the smoke that the crew had to wear gas masks for 2.5 hours afterwards. Slowly the fire was brought under control and finally extinguished. Jerry Linenger, the only doctor on board, carried out medical examinations on all the crew, knowing full well that smoke inhalation could have had serious repercussions.

The following morning Linenger asked Valeri Korzun for permission to speak to his NASA controller in the Russian Mission Control, so that he could get information regarding the toxic materials that could have been affected by the fire. Korzun kept promising that he could, but every time there was a communications pass Korzun was talking to the ground and would not let Linenger near the radio. Linenger got angrier and angrier when he realised that Korzun was not mentioning the fire, and the reason he wasn't being allowed to talk to his NASA contact was that NASA or the press had not been told of the fire and the Russians had no intention of informing them. There followed a blasing row between the two men which was only calmed down by Sasha Kaleri intervening. The atmosphere, in both senses of the word, went downhill rapidly. One week later, the carbon dioxide removal system failed and it was a couple of days before the next Progress M-33 supply spacecraft arrived with repair equipment and replacement oxygen-generating 'candles'.

Over the next three years, a further twenty-three shuttle flights took place: STS-70; 69; 73; 74; 72; 75; 76; 77; 78; 79; 80; 81; 82; 83; 84; 94; 85; 86; 87; 89; 90; 91; and 95, taking crews and supplies to the Mir Space Station as well as deploying satellites and carrying out experiments.

CHAPTER NINE

THE BUILDING OF THE ISS

The problems that were now being experienced aboard the ageing Mir Space Station were forcing the Americans and Russians to plan together and use each others' experience and expertise. Five years later, on 6 December 1998, STS-88 – Endeavour (OV-105) was launched and the building of the International Space Station got underway.

The crew of Robert D. Cabana – Commander; Frederick W. Stuckow – Pilot; Nancy J. Currie – MS; Jerry L. Ross – MS; James H. Newman – MS; and Russian Sergei K. Krikalyov – MS had one primary objective on this mission: to start the assembly of the International Space Station (ISS). This first section was made up of two modules: Zarya, which was owned by NASA but had been built by Russia, and Node 1 – Unity, which was built and owned by NASA. The two modules were connected by PMA-1 (Pressurised Mating Adaptor), a docking module. On the morning of 5 December Endeavour's docking system was connected to the 12.8-ton Node 1 module using PMA-2. The following day, the Remote Manipulator System arm was used to grasp the Zarya module from orbit, which was then mated to Unity. There were some initial problems when it was discovered that the Unity/Zarya fittings would not align properly and they had to be separated for adjustments to be made. Astronauts Jerry Ross and James Newman then commenced the first of three EVAs to attach power cables, connectors and handrails. With all the cables and connectors in place the modules were powered up. Problems again were encountered when some of the floodlights failed, plunging areas of the modules into complete blackness, but these were soon rectified. The shuttle returned to Earth on 12 December.

On 27 May 1999 STS-96 – Discovery (OV-103) lifted off the pad at the Kennedy Space Center with cargo for the ISS. After reaching orbit the Orbiter Discovery rendezvoused with the ISS and carried out docking procedures with PMA-1, the docking unit. Amongst the crew of Kent V. Rominger – Commander, Rick D. Husband – Pilot, Ellen Ochoa – MS, Tamara E. Jernigan – MS and Daniel T. Barry – MS, were Canadian Julie Payette – MS and Russian

The International Space Station in orbit around the Earth.

Valery I. Tokarov – MS. The following day astronauts Tamara Jernigan and Daniel Barry entered the tunnel adaptor hatch into the payload bay and during their 7-hour and 55-minute EVA transferred equipment to the exterior of the ISS. This consisted of the Integrated Cargo Carrier (ICC) which carried the Russian cargo crane Strela, mounted on the exterior of Zarya, the Russian section of the station; the US-built crane the ORU Transfer Device (OTDo); and SpaceHab Oceaneering Space System Box (SHOSS).

On 31 May the hatch to the Unity module was opened and the crew began to transfer equipment from the Orbiter to the module. This consisted of communications equipment and battery units. Once this was completed the crew then fitted sound insulation to the Zarya module. It was to be another three days before the transfer of equipment was completed, and on 3 June the Orbiter Discovery undocked from the ISS.

With the first sections of the ISS assembled and all necessary equipment installed, the Russians decided that it was time to say goodbye to Mir. On 27 August 1999, the Space Station was powered down and then placed into a free drift mode. At 1812 hours GMT the same day the hatch was closed for the last time by the three remaining cosmonauts, Viktor Afanasyev, Sergei Avdeyev and Jean-Pierre Haignere, who landed in Kazakhstan at 0035 GMT on 28 August 1999. For the first time since September 1989 there were no human beings in space.

The Space Shuttle programme continued to deliver satellites and carry out experiments whilst plans were being drawn up to expand the ISS and arrange for crews to be installed. Shuttle flights STS-93, 103, 99, 101 and 106 carried out flights to the ISS taking sections of the Space Station to be attached.

On 4 April 2000 the spacecraft Soyuz TM-30 blasted off the launch pad at Baikonur to rendezvous with the deserted Mir in an attempt to carry out repairs on a suspected air leak. The crew, Commander Sergei Zalyotin and Engineer Alexsandr Kaleri were to spend the next three months making the Space Station habitable again. The flight had been financed by a consortium of westerners to the tune of $20 million who have thoughts of turning the now defunct Space Station into the world's first space hotel. The two cosmonauts docked with Mir on 6 April 2000. After carrying out necessary maintenance and repairing a small air leak, the crew used their Soyuz spacecraft to raise the station's orbital height. Supply spacecraft were then sent so that the station could be restocked and then on 16 June the spacecraft undocked from the Space Station and returned to Earth. Despite all the work reactivating Mir it was decided that it was not a viable proposition and the Space Station was shut down on 16 June, and was left to de-orbit and burn up on re-entry.

The next visit to the ISS was on 12 October when the STS-92 – Discovery (OV-103) was launched. The crew of Brian Duffy – Commander, Pamela Melroy – Pilot, Leroy Chiao – MS, Peter Wishoff – MS, Michael Lopez-Algeria – MS and Koichi Wakata – MS were to carry out some of the most intensive work on a Space Station to date. The primary mission for this flight was to take the exterior framework, Z1 (Zenit), for the first US solar arrays to Unity on the ISS. The arrays, when fitted, would give early power to sections 4A and B on the Space Station. Later, when the US laboratory was installed, the Z1, which had four large gyroscopic devices within it called CMG (Control Movement Gyros), would be used to manoeuvre the ISS into its proper orbit.

The fitting of the PMA-3 (Pressurised Mating Adaptor-3) was the second of the objectives carried out and would provide an additional shuttle docking port. This was a necessary addition because as the ISS developed and grew even larger, the frequency of flights to the Space Station was going to increase.

The mission was planned to last twelve days, seven of which were to be spent docked with the Space Station. This included four EVAs and two ingress missions into the Space Station itself. Discovery returned to Earth on 24 October.

One week later, on 31 October 2000, the first crew to man the ISS arrived aboard Soyuz TM-31, comprising Commander William Shepherd and engineers Yuri Gidzenko and Sergei Krikalyov. They were known as the Expedition 1 crew. The Russian spacecraft docked at the Zvezda port on the ISS. After transferring the crew and supplies, the spacecraft undocked and moved around the ISS to the other side, freeing the port for the Progress unmanned supply spacecraft to dock at. This crew were to remain aboard the ISS for a number of months carrying out numerous experiments and general maintenance of the Space Station. Their spacecraft Soyuz TM-31 remained connected to the Space Station as their rescue vehicle.

The Expedition 1 crew of William Shepherd, Commander (A), and engineers Yuri Gidzenko (R) and Sergei Krikalyov (R).

The mission for STS-98 – Atlantis (OV-104) was to deliver a series of scientific experiments to the ISS and recover a number of completed ones. They also delivered over a ton of food and fuel to the Space Station. The primary objective of the mission was to deliver and permanently attach the second of the pressurised US modules, the space laboratory Destiny, to the Space Station, and in doing so expand the station's life support and attitude control capabilities.

The on-board Expedition 1 crew then moved the rescue vehicle, Soyuz TM-31, from the aft port to the Zarya Nadir port in a manoeuvre that took 1 hour and 31 minutes. The Orbiter Atlantis returned to Earth on 20 February 2001 after another successful mission to the ISS.

One month later the next Space Shuttle lifted off the pad at the Kennedy Space Center. STS-102 – Discovery (OV-103) was the eighth shuttle mission to visit the International Space Station and served as a crew rotation flight by delivering the Expedition 2 crew of James Voss – MS, Susan Helms – MS and Yuri Usachyov – engineer. The mission was launched on 8 March and was to last twelve days. It carried two Mission Specialists: Paul Richards and Andrew Thomas, whose primary objective was to connect the Leonardo Multipurpose Logistics Module (MPLM) that was being carried in the Orbiter's payload bay.

After docking with the ISS a number of experiments were carried out concerning the additional payloads carried aboard the spacecraft and repairs were completed out on a satellite. The mission also provided transportation back to Earth for three crew members that had been undertaking experiments on the Space Station. Two Americans and one Russian replaced the incumbent three crew members William Shepherd, Yuri Gidzenko and Sergei Krikalyov, who had arrived aboard Soyuz TM-31.

The flights to the ISS were becoming increasingly regular as the benefits of having a fully operational Space Station began to produce more and more scientific results. The arrival of STS-100 – Endeavour (OV-105) on 19 April brought more equipment and supplies. The main function of STS-100 was to service the ISS and deliver three robotic components to be installed during two spacewalks. The Canadian robotic arm CandArm 2 was fitted to the platform that had been installed by the previous crew. The crews were now becoming international with an Italian and Russian on board as Mission Specialists. The Rafaello logistics module was attached to the US laboratory Destiny, and then four days later returned to the cargo bay of Endeavour.

The flight of Soyuz TM-32 on 28 April 2001 provided the station with a new lifeboat. This flight also caused a great deal of controversy because the Russians took paying American tourist Dennis Tito aboard their spacecraft, which the Americans objected to on the grounds that the ISS was a scientific Space Station not a tourist hotel. It has to be said that the millions of dollars Tito paid for the flight helped the Russian Space Agency's sadly depleted funds. The crew of TM-32 transferred their contoured seats to Soyuz TM-31, leaving TM-32 as the ISS rescue vehicle, and returned to Earth in Soyuz TM-31.

Just two months after STS-100 had visited the ISS with additional equipment, the shuttle STS-104 – Atlantis (OV-104) arrived on the 14 July with more pieces of the station.

STS-102 crew patch.

Quest Joint Airlock about to be attached to the ISS.

The primary mission of STS-104 – Atlantis was to deliver and install the 6-ton Quest Joint Airlock on to the Unity module. The airlock consisted of two cylinders of 4m diameter and 6m in length. It was pressurised by high-pressure oxygen/nitrogen tanks that were mounted externally and was designed to be the single point of entry/exit for all future EVAs. When in position, the airlock would allow all Extra Vehicular Activities (EVA) to take place with each astronaut/cosmonaut wearing their own EVA suits. Up to this point, all personnel had to wear Russian EVA suits as the only entry/exit was in the Russian module connected to the ISS. The Leonardo logistics module was once again attached to the US laboratory Destiny, recovered six days later and returned to the cargo bay of Atlantis.

STS-105 – Discovery (OV-103), which arrived at the ISS on 12 August, was one of many straightforward transport missions, taking the Expedition 3 crew of Frank Culbertson – MS, and engineers Vladimir Dezhurov and Mikhail Tyurin to the ISS. With them came a number of scientific instruments and experiments. The Expedition 2 crew of James Voss – MS, Susan Helms – MS and Yuri Usachyov – Engineer, who had all been on the Space Station for five months, returned with the spacecraft.

On 21 October Soyuz TM-33, carrying two Russian cosmonauts, Viktor Afanasyev and Konstantin Kozeyev, and one French cosmonaut Claudie Andre-Dehays, was launched to the International Space Station. The crew returned in Soyuz TM-32 spacecraft after spending eight days on the ISS. The Soyuz TM-33 was to remain attached to the Space Station as a 'lifeboat' for the long-term crew that was remaining on the ISS.

Another of the delivery flights was launched from the Kennedy Space Center when STS-108 – Endeavour (OV-105) lifted off on 5 December for an eleven-day mission to the ISS.

The primary objective was to take the Expedition 4 crew of Yuri Onufrienko – Engineer (R), Daniel Bursch – MS, Carl Walz – MS to the Space Station and bring the Expedition 3 crew, Frank Culbertson – MS, Vladimir Dezhurov – Engineer (R), Mikhail Tyurin – Engineer (R), back. Shuttles were becoming more and more like delivery and refuse trucks when after unloading supplies the crews loaded up all the unwanted material and shipped it back to Earth. Once again the Rafaello logistics module was attached to the US laboratory Destiny and four days later, after numerous experiments, returned to the cargo bay of Endeavour. The spacecraft returned to Earth on 16 December after an almost faultless mission.

STS-109 – Columbia (OV-102) launched from the Kennedy Space Center on 1 March. It was the 108th mission of the Space Shuttle programme, the the twenty-seventh flight of the Orbiter Columbia, and the fourth servicing of the Hubble Space Telescope. It was also the last successful mission of the Orbiter Columbia before the ill-fated STS-107 mission, which became known as the Columbia Disaster.

The primary function of the mission was to service and modify the Hubble Telescope after which the discovery power of the telescope would be increased by the power of ten. The STS-109 astronauts performed a total of five spacewalks in five consecutive days to service and upgrade the Hubble Space Telescope. The spacewalkers received assistance from their crewmates inside Columbia. Nancy Currie operated the shuttle's robot arm while Scott Altman was her backup.

The Hubble Space Telescope (HST) had been placed in orbit during mission STS-31 on 25 April 1990. Initially it was designed to operate for fifteen years, but plans for periodic service and refurbishment were incorporated into its mission from the start. After the successful completion of the second planned service mission (SM2) by the crew of STS-82 in February 1997, three of HST's six gyroscopes failed. NASA decided to split the third planned service mission into two parts, SM3A and SM3B. A fifth and final servicing mission (SM4) was targeted for launch in May 2009. The work performed during SM4 is expected to keep HST in operation through to 2014. Further plans for servicing after SM4 are ambiguous, as NASA is planning to launch HST's successor, the James Webb Space Telescope in 2013.

Other spacewalk achievements included the installation of new solar arrays, a new camera, a new power control unit, a reaction wheel assembly and an experimental cooling system for the NICMOS unit. STS-109 accumulated a total of 35 hours and 55 minutes of EVA time. Following STS-109, a total of

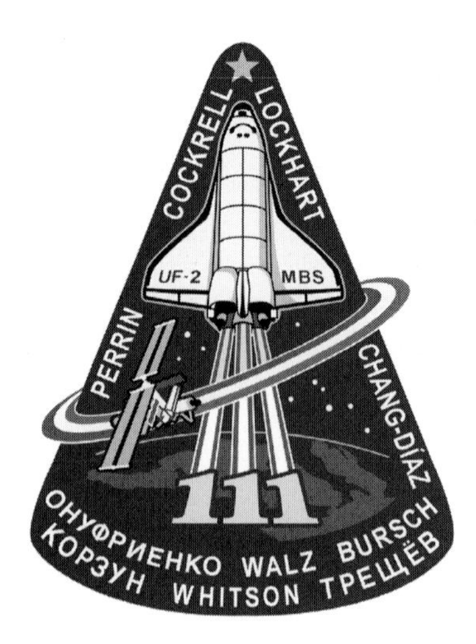

STS-111 crew patch.

eighteen spacewalks had been conducted during four Space Shuttle missions to service Hubble.

The development of the ISS continued with the delivery of the SO Truss by the STS-110 – Atlantis (OV-104) on 10 April. The crew, consisting of Michael Bloomfield – Commander; Stephen Frick – Pilot; and Mission Specialists Jerry Ross, Steven Smith, Ellen Ochoa, Lee Morin and Rex Walheim, had only one objective: to assemble and deliver a 91m station truss, known as the SO Truss, to be attached to the US laboratory on the ISS. This truss created a movable base for the CandArm 2 that allowed it to move along the station truss.

The SO Truss (also called the Centre Integrated Truss Assembly Starboard O Truss) forms the central backbone of the Space Station. It was attached on the top of the Destiny Laboratory Module during and used to route power to the pressurised station modules and conduct heat away from the modules to the S1 and P1 Trusses. The SO Truss is not docked to the ISS, but is connected with four Module to Truss Structure (MTS) struts.

Another 'space tourist' visited the ISS on 25 April 2002, aboard the Russian spacecraft Soyuz TM-34. South African internet millionaire Mark Shuttleworth paid $20 million for the trip. He was placed aboard the spacecraft, together with Italian cosmonaut Roberto Vettori and Russian cosmonaut Yuri Gidzenko, as a Mission Specialist. Unlike the Dennis Tito trip, the inclusion of Shuttleworth into the ISS crew was given approval by NASA because they now realised that it was one way for the Russians to boost their funding of the ISS. Their only reservations were about bringing untrained personnel into space, but Shuttleworth spent eight months training at Star City with other cosmonauts. The spacecraft returned to Earth on 5 May after spending ten days docked with the ISS.

Two months after the visit of the Russian spacecraft, the STS-111 – Endeavour (OV-105) arrived with the Expedition 5 crew: engineers Valeri Korzun and Sergei Treschev and Mission Specialist Peggy Whitson.

In addition to rotating the ISS crews, this mission was to carry out servicing on the ISS and replace the wrist roll joint on CandArm 2. The remote manipulator arm on Endeavour was used to remove the logistics module Leonardo from the spacecraft cargo bay and attach it to the Common Berthing Mechanism on the underside of Destiny. Chang-Diaz and Perrin carried out a spacewalk to check the thermal coverings. A second spacewalk by the two astronauts was carried out to attach video, data and power cables to the Mobile Base Unit. They also re-located one of the external television cameras. The wrist-roll joint on the CandArm 2 was replaced during the third spacewalk.

The Leonardo logistics module was removed from the cargo bay of Endeavour and attached to Destiny. Four days later it was returned to Endeavour's cargo bay and the spacecraft, with the Expedition 4 crew aboard, Mission Specialists Carl Walz and Daniel Bursch and engineer Yuri Onufriyenko, departed the ISS and returned to Earth on 19 June.

Shuttle flights continued to place satellites in orbit and carry out numerous medical and scientific experiments. Although vital and informative to the medical and scientific communities, these were of less interest to the general public. It was the visits to the ISS that helped capture the public's imagination as more and more women astronauts were taking their place aboard the ISS.

The launch of STS-113 – Endeavour (OV-105) on 23 November took the Expedition 6 crew to the ISS. The ongoing ISS crew consisted of Ken Bowersox, Donald Petit and Nikolai Budarin of Russia. On docking with the Space Station, supplies were transferred and a number of experiments carried out, including the fitting of an integral truss and spacewalk work platform on the port side of the ISS. This would support an additional cooling system which would be needed when the solar array panels were extended to almost an acre. To enable the ISS crews to work more easily outside the Space Station, human-powered railcars called Crew and Equipment Translation Aid (CETA) carts would be installed, to allow the EVA crews to traverse along the railway on top of the truss.

The returning Expedition 5 crew of Valeri Korzun and Segei Treschev of Russia, and Peggy Whitson embarked on the Endeavour for the return to Earth after having been on the ISS since 10 June. The eleven-day mission was a complete success.

The next Russian launch Soyuz TMA-1 on 30 October 2002 was the subject of some controversy. A problem arose when the American Lance Bass, of the pop band N Sync, was asked to leave Star City because of his lack of dedication to the training and the fact that he was unable to come up with the $13 million required. The Russians decided that he was too big a risk and dropped him from the mission to the ISS. Even when the crew went to the United States for

Crew of Soyuz TMA-1.

familiarisation training on the American segment of the ISS, they refused to take Bass with them, saying that he could go if he paid his own way and got permission from NASA. Suffice to say he pulled out of the project. He did, however, get a 'Cosmonaut's Certificate' for the training he had completed. He had been so confident in going that he had even had his name put on the crew patch. Bass was not the only 'space tourist' to apply at the time: Polish businessman Leszek Czarnecki had also applied, but when his plans had been leaked to the press he pulled out due to financial pressure.

Yuri Lonchakov took Lance Bass' place and joined Sergei Zalyotin and Frank de Winne. The launch of the Soyuz TMA-1 spacecraft was a modified, updated version of the TM-34 Soyuz spacecraft and took place on 30 October from Baikonur Cosmodrome amidst thick fog. The spacecraft docked with the ISS on 1 November and eight days later joined the long-term ISS crew of Kenneth Bowersox, Nikolai Budarin and Don Pettit.

De Winne carried out a programme of twenty-three experiments covering life and physical sciences and, after completing their experiments, the crew undocked and returned aboard the Soyuz TM-34 spacecraft, leaving the TMA-1 as the 'lifeboat' for the resident crew still aboard the ISS. These TMA flights were regarded as 'taxi' flights to deliver supplies and a fresh spacecraft.

On landing back on Earth the crew found themselves 460km off course and searches and rescue aircraft were sent to look for them. Two hours later they were spotted by a search aircraft, but it was to be a further 3 hours before rescue helicopters reached them. Initially the blame was placed on the cosmonauts for the spacecraft's deviation from the proposed landing site, or as one of the engineers put it, 'they pressed the wrong button', but later it was discovered that a fault in the spacecraft's computerised landing system was to blame.

In September the Chilean government entered into talks with the Russians with regards to sending one of their air force pilots into space. Their candidate, Klaus von Storch, arrived in Moscow for a series of medicals to determine whether or not he was fit for a spaceflight. After successfully completing the medicals, he was assigned for training for a flight aboard the Soyuz TMA-2 spacecraft. Unfortunately the money ran out and despite the Russian government's help, the Chileans could not raise or justify the millions required and so the dream faded.

Plans were put into action to replace the Chilean candidate with an experienced trained cosmonaut, Alexander Kaleri, but then all these plans were turned upside down with the devastating loss of STS-107.

The tragedy happened on 1 February 2003 when STS-107 – Columbia (OV-102) disintegrated on re-entry. Launched on 16 January, STS-107 went into orbit around the Earth at an altitude of 278km. After docking with the ISS and completing over eighty experiments as well as servicing the Space Station, Columbia undocked and headed for Earth. One of the experiments carried out by Ilna Ramon was to measure the impact of aerosols on cloud formation and rainfall, and the examination of red and blue flashes of light that appear during lightning storms.

On 1 February at 0800 EST, Columbia re-entered the atmosphere over Texas at a height of over 207,000ft and at a speed of 12,500mph. The first indications that anything was wrong came at 0753 hours when Flight Controllers detected a sudden temperature rise of 60°F in the fuselage above the left wing followed by a loss in the hydraulic system temperature measurements associated with the spacecraft's left wing. Three minutes later there was an increase in temperature on the left gear tyres and brakes. One minute later there was a loss of bondline temperature sensor data from the area of the left wing followed almost immediately by a loss of data and temperature and pressure readings from the left inboard and outboard tyres. It then appeared that a part of the Orbiter broke away during the descent causing the spacecraft to spin violently. Columbia was lost when it tore apart and exploded killing all seven crew members on board: Commander Rick Husband; Pilot William McCool and Mission Specialists Michael Anderson, Kalpana Chawla, Curtis Brown, Laurel Clark and Ilna Ramon, an Israeli.

The official report concluded that at 81.7 seconds after lift-off from pad 39A, a suitcase-sised chunk of foam broke away from the ship's external fuel tank and slammed into the left wing. The shuttle was racing skyward at more than twice the speed of sound at the time –1,650mph – and engineers later calculated the foam hit the left wing at some 530mph.

The foam strike was not seen until the day after launch when engineers began reviewing tracking camera footage as they do after every launching. A film camera in Cocoa Beach that could have photographed the impact on the underside of the left wing was out of focus. A video camera at the same site was properly focused, but it lacked the resolution or clarity to show exactly where the foam hit or whether it caused any damage. A third camera at a different site showed the foam disappearing under the left wing and emerging as a cloud of debris after striking the underside. Again, the exact impact point could not be seen.

Engineers immediately began analysing the available film and video, and ultimately determined that the foam had struck the heatshield tiles on the underside of the wing, near the left main landing gear door. No one ever seriously considered a direct hit on the reinforced carbon carbon panels making up the wing leading edge because no trace had been seen. It is difficult to understand why the impact on the leading edge did not receive more attention by the Ground Controllers and why they had not made the crew aware of their concerns.

It is thought that if there had been damage to the wing some of the damaged insulation tiles could have been lost as the spacecraft came back through the atmosphere, and the wing had overheated to the point of disintegration.

Within hours of the accident, the Russians launched a Progress unmanned supply spacecraft to deliver a ton of food, equipment and mail to the crew still aboard the International Space Station. The tragic accident meant that all shuttle flights were suspended and that Russian supply craft would have to make more trips to the ISS than had originally been planned. This in itself caused a number of problems as only two Russian TMA spacecraft capable of carrying cosmonauts/astronauts are built a year. The cost of building these spacecraft is extremely high and the Russian Space Agency is struggling to meet the cost of just these two, so to build more in a year was going to be extremely difficult unless additional funding was found. NASA approached the US Congress with a view to having their budget increased to help cover the cost of building Soyuz spacecraft, so that supplies and crews could be taken to the ISS, and an agreement was reached.

With the suspension of all the shuttle flights after the tragic accident, the only way to supply the ISS with crews and supplies was by means of Russian Soyuz spacecraft and unmanned Progress supply spacecraft. This tragic accident brought the two most powerful nations in spaceflight closer together and ensured an unprecedented degree of co-operation.

On 26 April TMA-2 was launched to relieve the crew aboard the International Space Station. The outgoing Expedition 7 crew of Yuri Malenchenko and Ed Lu had originally been three, but because of the tragic event the third member of the crew, Alexander Kaleri had been 'bumped'.

All flights to the Space Station had to be launched from Russia because of the hold on all shuttle flights in the United States after the tragedy of STS-107. The returning Expedition 6 crew of Mission Specialists Kenneth Bowersox and Don

Petit and engineer Nikolai Budarin returned to Earth aboard Soyuz TMA-1. Both members of the ongoing Expedition 7 crew were experienced cosmonaut/astronauts; Malenchenko had spent four months aboard the Russian Space Station Mir in 1994 and was aboard STS-106 when it docked with the ISS for the first time. Ed Lu was also on that flight, so the two were not strangers, which helped considerably when it was realised that between the two of them they were going to have to continue to operate the science payloads already on board, and maintain the Space Station.

During on the Space Station, Malenchenko became the first person to be married whilst in space. His bride, Ekaterina Dmitrieva, was in Texas at the time, where long-distance marriages are legal.

China launched its first manned spacecraft into orbit around the Earth on 15 October 2003. Thirty-eight-year-old Lieutenant-Colonel Yang Liwei was blasted into space in his Shenzhou 5 spacecraft by a Long March 2F rocket from the Jiuquan launch site in a remote part of the Gobi Desert in Gansu.

The spacecraft, very similar in design to the Russian Soyuz model, but with more advanced computer systems, made fourteen orbits of the Earth in 21.5 hours. Yang Liwei carried out a number of experiments, but no details of them are available. The spacecraft landed safely in Inner Mongolia. The launch and recovery were not broadcast live because had there been an accident the propaganda aspect of the mission would have been lost.

The Russians continued to deliver crews and supplies to the ISS and on 18 October, two days after the Chinese had carried out their first manned spaceflight, they launched Soyuz TMA-3. The ongoing Expedition 8 crew of Michael Foale, Alexander Kaleri and Pedro Duque, a visiting crew member, docked with the ISS two days later.

ESA astronaut Pedro Duque accompanied Michael Foale and Alexander Kaleri, who, after being 'bumped' from the previous flight, was back on flight status. They were to carry out normal housekeeping duties together with experiments, whilst Pedro Duque carried out his eight-day visit conducting experiments connected with the ESA's Cervantes Mission. He returned with the Expedition 7 crew of Yuri Malenchenko and Ed Lu.

The Soyuz flights to the ISS were fast becoming a cause for concern amongst both the RKK Energia and NASA. The whole space programme in Russia was being run on a shoestring and money was running out. NASA wanted to divert some of its finance in an effort to ease the situation, but the US Congress had prohibited the transfer of money to the Russians because of their involvement in helping Iran develop a nuclear capacity.

Russian space officials were becoming increasingly concerned regarding the future of the ISS, as funding became more difficult to send crews to man the

Soyuz TMA-3 approaching the ISS.

Space Station. It was because of this problem that increasing interest was being shown by the Russians in taking paying 'space tourists' to the ISS, despite reservations being voiced by NASA. NASA's main concern, on the other hand, was that, because of the desperate financial situation, corners were being cut in the construction of the Soyuz spacecraft to the detriment of safety.

Soyuz TMA-4 was launched by a Soyuz-FG rocket from Baikonur at 0319 UT on 19 April 2004. It carried the Expedition 9 crew of Michael Finke – MS; Gennady Padalka – Engineer; and Andre Kuipers – a visiting MS from the Netherlands – to the ISS and docked with the Zarya module automatically on 21 April at 0501 UT. Two of its astronauts, Padalka and Finke, remained in the ISS for about six months. The Dutch astronaut and the two astronauts who had inhabited the ISS for several months left the ISS on 29 April in the TMA-3 that had remained docked with the ISS, soft landing in Kazakhstan at 0011 on 30 April.

The continuous manning of the ISS was one of the most important factors in the space programme. This was highlighted by Gennadi Padalka who, when asked about the role of the astronaut on board the Space Station during an interview, said, 'We discovered an air leak in one of the outposts, which, had there not been a crew aboard to repair it, meant we could have lost the Space Station'.

During their mission the two crew members carried out a simultaneous spacewalk leaving, for the first time, an unattended Space Station. Procedures covering the spacewalks had been worked out previously between the American and Russian controllers on the ground. Two spacewalks had been planned, but a third was carried out when an external gyroscope component on the US section became faulty. A problem then arose with the American spacesuit, making it unusable. The Ground Controllers and the two crew members got together and worked out a method whereby Michael Finke could use the Russian spacesuit. It was successful, highlighting another area in which co-operation between the two countries was seen to be working.

After the problem had been identified and Michael Finke got to work and repaired the faulty American spacesuit. Usually in a case like this, the faulty spacesuit would be exchanged for a new one when the next shuttle arrived. But because of the suspension of the shuttle programme, the crew had to rely on unmanned Progress spacecraft for supplies, and there was no was no room for a new spacesuit on board. The Expedition 8 crew returned to Earth aboard the TMA-4 spacecraft.

The crew of Expedition 10 launched aboard Soyuz TMA-5 from Baikonur on 14 October, just two days behind schedule. The routine launch suddenly turned into a minor drama when, on the 16th, the autopilot that was to guide

The Soyuz TMA-4 crew.

Soyuz TMA-6 spacecraft approaching the ISS.

the spacecraft through the approach and docking procedure malfunctioned as it approached the ISS. Salizhan Shapirov immediately took over manual control and then guided the Soyuz spacecraft into the Pirs module-docking port. This confirmed that the hours spent in the Soyuz simulator practising manual control of the spacecraft had not been wasted.

It wasn't until the Expedition 9 crew had left on the TMA-4 spacecraft, which had been connected to the Zarya module, that the ongoing crew realised that the appetite of the returning crew had diminished the food supplies somewhat, and that Ground Control had not been informed of the problem. The Expedition 10 crew had only brought a limited supply of additional food, so they had to ration themselves until 25 December when a Russian Progress M-51 unmanned spacecraft docked with the ISS. On board was 1,234lb of propellant, 110lb of oxygen, 926lb of water and 2,777lb of dry goods, which included seventy food containers (forty-one American and twenty-nine Russian). If for some reason the Progress spacecraft had been unable to dock, then Leroy Chiao and Salizhan Sharipov would have had to return to Earth at the beginning of January. The crews on the ISS were becoming totally reliant upon the Russian Progress unmanned cargo spacecraft for supplies.

On 15 April 2005 the Expedition 11 crew was launched from Baikonur aboard Soyuz TMA-6, to relieve Leroy Chiao and Salizhan Sharipov. The crew, consisting of Sergei Krikalev, John L. Phillips and Italian Roberto Vittori went aboard the Space Station on 16 April. Vittori returned to Earth aboard the Russian spacecraft TMA-5 on 24 April with the Expedition 10 crew after completing a number of experiments for the European Space Agency. A third member of the Expedition 11 crew, German Thomas Reiter, was scheduled to join Krikalev and Phillips when the Space Shuttle STS-121 was launched to re-supply the ISS.

There were one or two worrying problems for Chiao and Sharipov during their six-and-a-half-month stay aboard the ISS, notably when one of the main gyros that kept the ISS in a stable orbit malfunctioned. Fortunately, the two remaining gyros managed to keep the Space Station steady whilst the third gyro was repaired. The majority of the tasks carried out by the crew were the normal housekeeping duties and the constant maintenance needed to keep the Space Station up and running.

On 17 June the unmanned supply spacecraft Progress 18 lifted off the pad at Baikonur with supplies for the ISS. Everything went well until the final docking commands were needed and then the Ground Controllers found the spacecraft was not responding. Sergei Krikalev immediately took over control and eased the spacecraft into one of the docking ports in the Russian Zvezda Service Module.

The following day Sergei Krikalev and John Phillips opened the hatch in Progress 18 and removed the supplies. The previous supply spacecraft, Progress 17, had been filled with trash, waste and items no longer needed, and then released to burn up in the atmosphere on re-entry. The supplies carried aboard the spacecraft amounted to 4,662lb, of which 3,100lb were dry cargo such as food and experimental equipment.

STS 114 – Discovery (OV-103) was the first Space Shuttle flight for two years and lifted off the pad at Cape Canaveral on 26 July 2005. The mission was to re-supply the ISS and to test the improvements made to the Orbiter. A problem was noticed during lift-off, when the main fuel tank had a bird strike, and as the shuttle roared upwards, a piece of foam was seen to fly off the fuel tank. Since the last tragic flight, cameras had been mounted strategically to enable Ground Control to monitor the launch from a different perspective.

The crew of Eileen Collins – Commander; James Kelly – Pilot; and Mission Specialists Soichi Noguchi, Stephen Robinson, Andrew Thomas, Wendy Lawrence and Charles Carmada, attempted several new techniques during this mission. Among them was a repair to the Orbiter's heatshield. This was a vital experiment because it would enable astronauts to carry out repair work to the spacecraft's external skin in the event there was a problem. During the mission, Stephen Robinson and Soichi Noguchi had to carry out three EVAs in order to repair the Orbiter's heatshield. The work was a complete success.

On the fourth day in space, the Italian-built Multi-purpose Logistics Module (MPLM), also known as Rafaello, was lifted out of the cargo bay by means of the CandArm 2 and attached to the ISS. This was the module's third trip to the Space Station. When this had been completed, work commenced on transferring several tons of supplies and water to the station.

Because of the tragedy that had struck the previous shuttle flight, it was decided to put a second shuttle on the pad in case a rescue mission had to be launched. STS-300 – Atlantis was the first of these and was on stand-by for missions STS-114 and STS- 121.

The launch of Soyuz TMA-7 carried the Expedition 12 crew of William McArthur – Commander, Valeri Tokarev – Engineer, and the third paying tourist, American Gregory Olsen. The multi-millionaire paid $20 million to the firm Space Adventures, who brokered the flights to the ISS. Unlike the previous two 'tourists', Olsen considered himself a scientist and carried out a number of experiments, including serving as a test subject for two human physiology studies for the ESA (European Space Agency), studying motion sickness and lower back pain. The latter is a condition that almost all astronauts suffer from with long duration spaceflights. Olsen returned with the Expedition 11 crew: Sergei Krikalev and John Phillips.

The remaining members of the crew started their six-month residence of the Space Station. As with all previous crews, most of their time would be spent keeping up with maintenance and housekeeping duties, interspersed with various experiments. One of these was to boost the Space Station into a higher orbit using the Progress 19 spacecraft's engines. The spacecraft was docked at the aft end of the Zvezda module and the idea was to carry out two burns, each to last 11 minutes and 40 seconds, that would raise the ISS from its orbital peak of 220 miles to 224 miles. But when the engines were fired they turned themselves off less than 2 minutes into the burn. Officials have put the problem down to a fault in the data transmission which was fixed later.

The second manned Chinese spaceflight took place on 12 October, when astronauts Fei Junglong (Commander) and Nie Haisheng (Pilot) were launched from the Jiuquan Space Centre in the Gobi Desert in their spacecraft Shenzhou 6. This time the launch of their Shenzhou 6 spacecraft was shown live on state television, and the country watched as it hurtled into space and then into orbit. The two astronauts planned to orbit the Earth for about five days and carry out numerous experiments. The Chinese government initially expressed an interest in seeing the space programme more as a military objective rather than a commercial one, but has since stated that they are opposed to deploying any weapons in space.

A slight problem occurred on the thirtieth orbit, when the spacecraft deviated from its orbit and had to be placed back manually by firing its thruster rockets. After five days in space the spacecraft landed safely in Inner Mongolia. The Shenzhou spacecraft is based on the three-man Russian Soyuz spacecraft, as are the spacesuits and other life-support systems.

The House of Representatives voted on 26 October 2005 to allow NASA to buy the Russian Soyuz spacecraft so that astronauts and supplies could to be delivered to the ISS. Since the Space Shuttle programme was shut down in

The Soyuz TMA-8 crew.

2003, following the Columbia accident, the Russian Soyuz spacecraft has been the only means of getting crews and supplies onto the Space Station. The decision also meant that the Russian space programme's financial problems would be eased considerably.

In future the Space Shuttle would be used to launch large satellites into orbit and other large objects to the ISS, whilst the Soyuz spacecraft would be used to transport crews on a regular basis.

On 29 March 2006 the first Brazilian to go into space, Lieutenant-Colonel Marcos Pontes, lifted off with the Expedition 13 crew from Baikonur aboard a Soyuz TMA-8 spacecraft. The new ISS crew of Pavel Vinogradov of Russia and Jeffrey Williams of the USA were on their way to relieve the Expedition 12 crew. The cost to the Brazilian Space Agency of sending Marcos Pontes to the ISS was in excess of $10 million.

After docking with the ISS the three crew members transferred to the Space Station to be greeted by the two crew members they were to relieve. Preparation began almost immediately to take over the running of the Space Station and, ten days later, take-over of the station was officially handed over to Pavel Vinogradov and Jeffrey Williams.

The Expedition 12 crew, together with Marcos Pontes, returned to Earth, landing on the steppes of Kazakhstan on 8 April. This was Bill McArthur's last spaceflight as he had decided to retire from the space programme after this, his fourth flight.

Jeffrey Williams inside the Soyuz TMA-8 spacecraft highlighting the extremely cramped conditions the cosmonauts endured.

In an effort to catch up with their work schedule, the two astronauts carried out an extended spacewalk on 2 June. Wearing Russian-built Orlan spacesuits they exited the Space Station and made a number of crucial repairs to its systems. This included replacing a clogged vent nozzle that was used to dump excess hydrogen produced by the orbital laboratory's Elektron oxygen generator. A broken video camera, used to provide views of the robotic arm that was attached to the station's main truss segment, was also replaced. This was necessary because later, astronauts from STS-121 installed new solar arrays outside the ISS.

The crew of the ISS awaited the arrival of Discovery on the 6 July with great interest. This would be the first contact they had had with other humans for three months and it was also bringing a third ISS crew member, Thomas Reiter of the European Space Agency (ESA). Reiter was scheduled to work alongside the Expedition 13 crew and then join the Expedition 14 crew when it arrived in September.

After two delays, STS-121 – Discovery (OV-103) lifted off the pad at Kennedy Space Center on 4 July 2006. The delays had been caused by a number of problems including the finding of cracks in the foam insulation surrounding the external tank. Also during turnaround operations, it was discovered that a 3in piece of foam, weighing less than one-tenth of an ounce, had detached itself. At first there were fears that ice could form inside the hole it had left, but after detailed examination no trace of ice, or indeed water, was found. With the problem resolved, engineers decided that it offered no threat to the crew or the mission and approved the launch. Two days after lift-off, the Orbiter Discovery docked with the ISS. Just prior to the docking the spacecraft rolled onto its back so that the crew of the Space Station could study the thermal tiles underneath the Orbiter to ensure there were no damaged ones. After docking, the shuttle crew of Steven Lindsey – Commander; Mark Kelly – Pilot; Michael Fossum – MS; Lisa Nowak – MS; Piers Sellers – MS; and Stephanie Wilson – MS, prepared to carry out the repairs to the tiles in the aft section of the Orbiter's payload bay, which had been deliberately damaged. Using a heat-resistant material called NOAX, the putty-like substance was spread over the cracks and gouges in the tiles using a spatula. After completing the experiment Piers Sellers noticed that the spatula had somehow drifted off into space. It joined the rest of the debris that was now floating above the Earth. This was becoming a source of concern for everyone within the astronaut/cosmonaut world, as the area of space above the Earth was becoming a veritable junkyard of discarded satellites and remnants of space material all operating in an orbit with active satellites. Even the smallest, insignificant piece of material or discarded equipment could be a serious threat to the crews aboard the ISS and the spacecraft shuttling backwards and forwards.

The main object of the mission was to embark Expedition 14 crew member Thomas Reiter – MS (Ger). He would be joined later by two more crew members brought to the ISS by the Russian spacecraft TMA-9.

STS-121 returned to Earth without any problems and reassured by the fact that they knew now they could affect repairs to the tiles whilst in space.

The mission of STS-115 – Atlantis (OV-104) was to resume the construction of the ISS by installing a 17.5-ton integrated truss segment to the station's girder backbone. Attached to the girder were a set of large solar arrays, electronics and batteries. This new segment provided one quarter of the total power generation required for the Space Station.

STS-121 crew patch.

The spacecraft docked with the ISS on 11 September after chasing the Space Station for two days, and within hours of docking the crew of Atlantis, Brent Jett – Commander, Christopher Ferguson – Pilot, Joseph Tanner – MS, Daniel Burbank – MS, Heidemarie Stefanyshyn-Piper – MS and Steven MacLean – MS (Can), were aboard the ISS greeting the on-board three-man crew. Another new innovation was the introduction of the 'campout pre-breathing exercise'. Crew members that were to carry out the EVAs slept in the station's airlock overnight, whilst the pressure was slowly reduced. This removed any harmful gases from their blood, allowing their bodies to gradually acclimatise to the lower pressure they experienced inside their spacesuits when outside the Space Station.

The following day the Orbiter's crew lifted the two new segments of the Space Station from Atlantis' cargo bay using the CandArm. The two astronauts who guided the segments out of the cargo bay were the experienced astronaut Joe Tanner and the novice astronaut Heidemarie Stefanyshyn-Piper, who was making her first spaceflight.

With the mission completed, the spacecraft Atlantis undocked from the ISS. Once the first segment was clear of the cargo bay the second pair of astronauts, Daniel Burbank and Steven MacLean, replaced the first pair, and after a 7-hour spacewalk connected the first segment to the Space Station.

After spending twelve days in space, Atlantis touched down safely at the Kennedy Space Center. The rescue shuttle Discovery was on the launch pad throughout the mission.

In Russia, plans were well under way for the next Soyuz mission, TMA-9, which would carry the next ISS Expedition 14 crew of Miguel Lopez-Algeria, Mikhail Tyurin and the Japanese businessman Daisuke Enomoto to the Space Station. Enomoto, who had paid $20 million, was to have been the fourth paying passenger to visit the Space Station, but on 21 August he was removed from the crew because of an undisclosed medical problem. Back-up entrepreneur Anousheh Ansari took his place.

Lopez-Algeria and Tyurin were both experienced space travellers. Tyurin had already spent 125 days aboard the ISS as a member of the Expedition 3 crew, whilst Lopez-Algeria had three spaceflights to his credit within a seven-year period.

This was a unique flight to the ISS inasmuch as it carried the first female space tourist – Anousheh Ansari. The Iranian-born American entrepreneur replaced the Japanese businessman, Daisuke Enomoto, who was forced to stand down because of a medical problem. Miguel Lopez-Algeria and Mikhail Tyurin were to take the place of Jeffrey Williams and Pavel Vinogradov, who returned to Earth ten days later with Anousheh Ansari. Thomas Reiter, who had been taken to the ISS by STS-121, remained until December when American astronaut Sunita Williams replaced him.

Launched on 18 September, the Soyuz TMA-9 spacecraft orbited the Earth for two days before catching up and docking with the ISS. Three hours after docking,

Soyuz TMA-9 crew, including the first female space tourist, Anousheh Ansari.

the hatches were opened and the replacement crew entered their new home to be welcomed by the on-board crew. Anousheh Ansari, although having paid an estimated $20 million for the flight, was also going to carry out a series of medical experiments for the ESA. In a very frank interview later, Ansari described the space sickness she experienced during the first two days of the mission. Normally, both the Americans and Russians gloss over this reaction by astronauts and cosmonauts to space sickness, but Ansari described how she suffered from severe back pain, headaches and nausea for two days; the only relief she got was from motion sickness injections which made her very sleepy. On reaching the Space Station, she felt 100 per cent, and described weightlessness as one of the most euphoric sensations she had ever experienced.

On 10 October the crew climbed aboard the Soyuz TMA-9 spacecraft, which was docked in the Zvezda Service Module, and moved it to the Zarya control module. This 'quick trip around the block', as Lopez-Algeria put it, was to free the docking port for the arrival of the Progress 23P unmanned supply spacecraft later that month. During the 20-minute flight, the crew wore the Russian Sokol spacesuits. After docking with the ISS, the crew spent 2 hours carrying out leak checks before opening the hatches into the Space Station.

Originally scheduled for launch on 7 December, but cancelled because of low cloud, was STS-116 – Discovery (OV-103). Its primary objectives were the delivery and attachment of a third port truss section for the ISS, the rewiring of the

STS-116 crew patch.

ISS power station and the transfer of an ISS crew member. Once again there was a rescue shuttle, Atlantis, on the launch pad for the duration of the mission if required.

Three days after the launch, Discovery docked with the ISS and the Orbiter's crew retracted one of the two 120ft solar arrays from the station's Port 6 in order to overhaul it. With this completed, the new Port 5 truss was then installed. Up to this point in time, all the power for the ISS had been provided by solar arrays on the Port 6 truss. The addition of the Port 3 and 4 trusses meant that more power could be provided for the additional sections that were going to be fitted to the ISS.

With the overhaul and the fitting of the Port 5 truss completed, Discovery separated from the ISS and made its now traditional fly-around so that photographs could be taken of the Space Station, showing the additional work that had been done. The crew of STS-116 then returned to Earth for a very welcome Christmas vacation.

Crew: Mark Polansky – Commander (A), William Oefelin – Pilot (A), Nicholas Patrick – MS (A), Robert Curbeam – MS (A), Arne Fuglesang – MS (Swed), Joan Higginbotham – MS (A).
Embarking Expedition 15 crew member: Sunita Williams (A).
Returning Expedition 14 crew member: Thomas Reiter (Ger).

With the launch of Soyuz TMA-10 on 7 April 2007, another space tourist got ready to visit the Space Station. American software developer Charles Simonyi, who had created a large number of programmes for Microsoft, paid $20 million for the privilege. He also got an additional two days on the station because of its orbital position and the delay in the launch of STS-117. During his stay on the station, like all the other space tourists before him, Simonyi had to carry out experiments for the space agencies. His task was to participate in a series of biomedical experiments that were carried out during his thirteen-day visit. The other two Expedition 15 crew members, Fyodor Yurchikhin – Commander (R) and Oleg Kotov – Engineer (R), were joining Sunita Williams who was already aboard the ISS, having been taken there aboard STS-116.

Charles Simonyi returned to Earth on Soyuz TMA-9 with the Expedition 14 crew members.

The primary mission of the crew of STS-117 – Atlantis (OV-104), Frederick Sturckow – Commander; Lee Archambault – Pilot; Patrick Forrester – MS; Steven Swanson – MS; James Olivas – MS; James Reilly – MS; and Clayton Anderson – MS, which was launched on 8 June 2007, was to deliver a 17½-ton truss segment, together with large solar wings that, when connected, stretched 240ft in length once opened out to their fullest extent. Their second mission was to deliver a replacement Expedition 16 crew member to the ISS – Clayton Anderson. Sunita Williams, who he was to replace, had been on the Space Station since December 2006. She then returned to Earth aboard STS-117.

The new solar array, known as S4, was to provide additional electricity for the new modules and laboratories that were scheduled to arrive in 2008. The S4's design allowed it to rotate like a paddlewheel so that it could follow the sun. On Monday 11 June Steve Swanson and Patrick Forrester carried out a spacewalk, firstly to retract an older solar array that was covering the area, and then to install the S4. With this mission completed, during the early hours of the following morning Mission Control unfurled the solar wings of S4 and by the early afternoon they had opened to their full extent. The need for the wings to open slowly was because the paper-thin photovoltaic cells which made up the array had a tendency to stick together when folded. By allowing the sun to warm them, this stopped the process known as 'stiction' from happening.

It was discovered during the launch that a 10 x 15cm triangular-shaped piece of thermal blanket on the port Orbital Manoeuvring System (OMS) pod had come loose. A spacewalk by Danny Olivas and James Reilly was carried out to repair the loose piece using a medical stapler with nickel chromium pins.

Then on 13 June a problem arose, when the six Russian computers that controlled the orientation and navigation systems for the ISS went off-line. This meant that the Russian-built thrusters that kept the Space Station oriented became inoperative. The Orbiter Atlantis' thrusters, together with American-built control movement gyroscopes, were used to keep the ISS oriented. The problem that was causing the most concern was that when Atlantis left the ISS to return to Earth, it was not known if the Space Station would be able to maintain its orientation and orbital path. One option was to use the thrusters on board the two

Progress cargo ships that were attached to the ISS, if the computer problem had not been resolved.

The computers also controlled the primary Elektron oxygen generator, which caused them to shut down at the same time. Fortunately, there was a sufficient supply of stored oxygen to support the crew for over a month and it was hoped that the computer problem would be resolved well before then.

The computer malfunction was finally resolved by installing jumper cables to bypass the faulty hardware and the Space Station was back on-line and self-supporting. The arrival of STS-118 would bring a new computer system to replace the ageing one that was causing problems.

Endeavour was the back-up shuttle throughout the mission.

Back on the ISS, the crew made preparations for the arrival of the unmanned Progress 26 supply craft. Laden with 1,600lb of propellant, 100lb of oxygen and air, and 496lb of water, in addition to the fresh clothing, food, mail and numerous scientific instruments, the spacecraft was a welcome relief to the humdrum regime of the crew. They packed the Progress 24 supply craft with all their unwanted rubbish and jettisoned it from the Piris module, leaving just the Progress 25 supply craft docked in the Zvezda module and the Soyuz TMA-10.

The STS-118 mission to the ISS, flown by Endeavour (OV-105), successfully lifted off from the Kennedy Space Center on 8 August 2007. This was the first flight of Endeavour since the STS-113 mission in November 2002, which was the last successful Space Shuttle flight before the loss of Columbia on STS-107. Had the Orbiter Columbia not disintegrated, it would have been chosen for this flight, which would have been its twenty-ninth mission, and probably its only mission to the ISS.

The crew of Scott Kelly – Commander; Charles Hobaugh – Pilot; Tracy Caldwell – MS; Richard Mastracchio – MS; Daffyd Williams – MS; Barbara Morgan – MS; and Benjamin Drew – MS, delivered and assembled the starboard S5 truss segment of the International Space Station, as well as External Stowage Platform 3 (ESP-3). The mission was also the final flight to include the Spacehab Logistics Single Module.

The Spacehab Logistics Single Module, a pressurised aluminum habitat that is carried inside the payload bay, has a capacity of 6,000lb (2,722kg), and carried a variety of cargo and research projects, including supply materials for the ISS. It returned with approximately 3,000lb (1,361kg) of cargo, including the MISSE PEC, a DoD payload that had been installed on the ISS. Launched in July 2006, the payload contained over 850 materials specimens that were to be studied to determine the effects of long-term exposure to the environment of space.

The spacecraft returned to Earth on 21 August landing at the Kennedy Space Center.

Discovery was the back-up shuttle.

STS-118 crew patch.

The launch of Soyuz TMA-11 on 10 October was the cause of some controversy when one of the crew was Sheikh Muszaphar, who flew as a guest of the Russian government. In exchange for the multi-billion dollar purchase of fighter jets by Malaysia, the Russian Federation bore the cost of training two Malaysians for space travel and for sending one to the ISS.

Sheikh Muszaphar's role aboard the Soyuz is referred to as a Spaceflight Participant by the Russian Federal Space Agency and NASA during briefings. Speaking to Malaysian media outlets, Alexander Karchava, the Russian ambassador to Malaysia, stated that Sheikh Muszaphar was a 'fully-fledged cosmonaut' after media criticism about the Malaysian being just a space tourist. Muszaphar returned aboard Soyuz TMA-10.

The prime object of the mission was to deliver two Expedition 16 crew members to the ISS, Yuri Malenchenko (R) and Peggy Whitson, to join Clayton Andrson who arrived aboard STS-117.

STS-120 – Endeavour (OV-103) delivered the US Harmony Module (also known as Node 2), with four DC-to-DC Converter Unit (DDCU) racks and three Zero-G Storage Racks (ZSR) installed; a Power and Data Grapple Fixture (PDGF) for the station's robot arm; and a Shuttle Power Distribution Unit (SPDU), to the International Space Station (ISS). Harmony was the first pressurised habitable module delivered to the station since the Pirs docking compartment was installed in August 2001.

The crew of STS-120, Pamela Melroy – Commander; George Zamka – Pilot; Douglas Wheelock – MS; Stephanie Wilson – MS; Scott Parazynski – MS; and Paolo Nespoli – MS ESA (Italy), docked to the Pressurised Mating Adaptor (PMA-2) on the forward port of the Destiny Laboratory Module. The installation of Harmony occurred in two stages. Firstly, the crew installed Harmony to the port node of the Unity module. After Discovery undocked, the station's robotic arm detached PMA-2 from Destiny and moved it to the forward port of Harmony. Following the relocation of PMA-2, the robotic arm had moved Harmony from its initial position to its final position on the forward port of Destiny.

The final positioning of Harmony allowed for the future installation of the ESA's Columbus and Japanese Kibo research modules, which were attached to the side ports of Harmony.

One of the other primary objectives of the mission was to deliver Expedition 16 crew member Daniel Tani to the ISS to replace fellow American Clayton Anderson, who returned to Earth aboard STS-120.

Atlantis was the back-up shuttle for this mission.

The launch of STS-122 Atlantis (OV-104) on 7 February 2007 was the ISS Assembly Flight 1E, which delivered the European Columbus laboratory module to the station along with two other medical laboratories. The Orbiter also carried the Solar Monitoring Observatory (SOLAR), a new Nitrogen Tank Assembly mounted in the cargo bay of an ICC-Lite payload rack, as well as a spare Drive Lock Assembly (DLA) sent to orbit in support of possible repairs to the starboard Solar Alpha Rotary Joint (SARJ), which was still malfunctioning.

Several items were returned with Atlantis: a malfunctioning Control Movement Gyroscope (CMG), which was swapped out with a new one during a visit by STS-118; and the empty Nitrogen Tank Assembly, placed in the Orbiter's payload bay along with a trundle bearing from the Starboard SARJ, that was removed during an EVA performed by Expedition 16.

It also delivered an ISS Expedition 16 crew member Leopold Eyharts (Fr) of the ESA to the Space Station as a replacement for Daniel Tani, returning Expedition 16 crew member.

Discovery was the back-up shuttle.

Crew: Stephen Frick – Commander (A); Alan Poindexter – Pilot (A); Leland Melvin – MS (A); Rex Walheim – MS (A); Hans Schlegel – MS (Ger); Stanley Love – MS (A).

STS-123 Endeavour (OV0-105) was a mission to the ISS to install the 1J/A. Launched on 13 March, this was the twenty-fifth shuttle mission to visit the ISS, and delivered the first module of the Japanese laboratory, Japanese Experiment Module (Kibo), and the Canadian Special Purpose Dexterous Manipulator (SPDM) robotics system to the station. The mission duration was 15 days and 18 hours, and it was the first mission to fully utilise the Station-to-Shuttle Power Transfer System (SSPTS), allowing Space Station power to augment the shuttle power systems. The mission also delivered Expedition 17 crew member Garrett

Reisman to the ISS, and returned Expedition 16 crew member Leopold Eyharts (Fr) back to Earth.

The mission set a record for a shuttle's longest stay at the ISS – fifteen days. The completion of the mission left nine flights remaining in the Space Shuttle programme until its end in 2010, excluding two as-yet-unconfirmed Contingency Logistic Flights.

Discovery was the back-up shuttle.

> Crew: Dominic Gorie – Commander (A); Gregory Johnson – Pilot (A); Robert Behnken – MS (A); Michael Foreman – MS (A); Takao Doi – MS (Jpn); Richard Linnehan – MS (A).

The Russians continued to send manned and unmanned spacecraft to the ISS. On 8 April they launched Soyuz TMA-12 from Baikonur on what was becoming a regular 'taxi' run to the Space Station. The crew consisted of Sergei Volkov (R), Oleg Kononenko (R) and Yi So-yeon of South Korea, the two Russians being the Embarking Expedition 17 crew.

Yi So-yeon flew as a guest of the Russian government through the Korean Astronaut programme, after the Korean government paid the Russian government US$25 million in agreement to support the first Korean astronaut in space. Her role aboard the Soyuz was once again referred to as a Spaceflight Participant by the Russian Federal Space Agency and NASA during press briefings. Ko San was originally scheduled to fly, with Yi as her backup. On 10 March 2008 it was

Crew of Soyuz TMA-12.

announced that Ko had breached regulations surrounding the removal of books from the training centre in Russia, and therefore was not allowed to fly.

The two Russian crew members replaced Peggy Whitson and Yuri Malenchenko who returned to Earth, together with Yi So-yeon, aboard the TMA-11 spacecraft on 19 April.

Launched on 31 May, STS-124 Discovery (OV-103) delivered the second part of the Pressurised Module (PM) of the Japanese Experiment Module (JEM), known as Kibo, to the International Space Station (ISS). Kibo was berthed to the Harmony module and the pressurised section of the JEM Experiment Logistics Module, brought up by the STS-113 crew, and was moved from Harmony to the JEM-PM. The crew of Mark Kelly – Commander, Kennet Ham – Pilot, Karen Nyberg – MS, Ronald Garan – MS, Michael Fossum – MS and Akihito Hoshide – MS (Jpn), also delivered the Japanese Remote Manipulator System, a robotic arm, and attached it to Kibo. The entire Kibo laboratory was being brought up over three subsquent missions.

The spacecraft Discovery carried replacement parts in a mid-deck locker for a malfunctioning toilet. The crew had been using other facilities for waste until the new replacement parts were installed on the Zvezda module of the ISS.

The spacecraft also delivered Expedition 18 crew member Gregory Chamitoff to the ISS, and collected Expedition 17 crew member Garrett Reisman and returned him to Earth. With the completion of the mission, Discovery returned to Earth on 14 June after spending almost fourteen days in space.

Endeavour was the back-up shuttle.

The spacecraft TMA-13 was launched on 12 October and its mission was to take the remaining two members of the Expedition 18 crew, Yuri Lonchakov – Commander (R) and Michael Finke – Flight Engineer, to the ISS. On board was also the latest fare-paying passenger, Richard Garriot, son of former astronaut Owen Garriot (Skylab 3 and STS-9), who returned to Earth on 12 October with the crew of TMA-12.

The last shuttle flight of 2008, STS-126 Endeavour (OV-105), was a mission to the ISS. The purpose of the mission, referred to as ULF2 by the ISS programme, was to deliver equipment and supplies to the station, to deliver the third member of Expedition 18, Sandra Magnus, and collect returning Expedition 17 crew member Gregory Chamitoff (A).

The shuttle crew of Chris Ferguson – Commander; Eric Boe – Pilot; Donald Petit – MS; Stephen Bowen – MS; Heidemarie Stefanyshyn-Piper – MS; and Robert Kimbrough – MS (A) also serviced the Solar Alpha Rotary Joints (SARJ), and repaired the problem in the starboard SARJ that had limited its use since STS-120.

STS-126 launched on 14 November 2008 from the Kennedy Space Center with no delays or issues. The spacecraft Endeavour successfully docked with the station on 16 November. This was scheduled to be a sixteen-day mission with four spacewalks, largely dedicated to servicing and repairing the SARJ. An additional docked day was added to the flight plan to give the crew more time to complete their tasks. Both the starboard and port SARJs were serviced. In addition to lubricating both bearings, the remaining eleven trundle bearings in the starboard SARJ were replaced.

STS-126 included the Leonardo Multi-purpose Logistics Module (MPLM) on its fifth spaceflight. Leonardo held over 14,000lb of supplies and equipment. Among the items packed into the MPLM were two new crew quarters racks, a second galley (kitchen) for the Destiny laboratory, a second Waste and Hygiene Compartment (lavatory) and a number of new experiments. Also included in Leonardo was the General Laboratory Active Cryogenic ISS Experiment Refrigerator (or GLACIER), a double locker cryogenic freezer for transporting and preserving science experiments.

After spending 11 days, 16 hours and 46 minutes docked to the station, during which the crew performed four spacewalks and transferred cargo, the Orbiter undocked on 28 November.

Discovery was the back-up shuttle.

On 25 September the Chinese launched a three-man spacecraft, Shenzhou VII, on top of a Long March 2F rocket, from their Jiuquan rocket base in the Gobi Desert. The three Taikonauts, as they are called, Zhai Zhigang, Liu Boming and Jing Haipeng, are lieutenant-colonels in the Chinese Air Force.

Two days later one of the three Taikonauts, Zhai Zhigang, wearing a Chinese-developed Feitian spacesuit, carried out a 20-minute spacewalk, the first ever for a Chinese astronaut. During this spacewalk, Zhai retrieved experiment samples and waved the Chinese flag in space. Cables were used to tie Zhai to the handrail outside the orbital module, and his moving route was restricted to areas near the exits. Liu Boming, wearing a Russian Orlan-M suit, stayed in the airlock in the orbital module to provide help if necessary. The third astronaut, Jing Haipeng, remained in the re-entry module to monitor the general situation of the spacecraft.

The Feitian spacesuit is similar to the Orlan-M in shape and volume, and is designed for spacewalks of up to 7 hours, providing oxygen and allowing for the excretion of bodily waste. Each suit was reported to have cost about $4.4 million. Except for the gloves of the Feitian suit, the spacesuits were not brought back to Earth.

STS-119 Discovery (OV-103) was a mission to the Space Station which was flown on 15 March 2009 to deliver and assemble the fourth starboard Intergrated Truss

Segment (S6), and deliver the fourth set of solar arrays and batteries. Discovery successfully landed on 28 March 2009, at 1513 hours EDT.

During the countdown, a bat was seen to be resting on the external tank. What was originally believed to be a fruit bat was revealed to be a free-tailed bat that clung onto the fuel tank during the launch. NASA observers had believed the bat would fly off once the shuttle started to launch, but it did not, and it was probably shaken off and incinerated by the rocket exhaust. A bat expert, analysing pictures, believed the bat had a broken wing which made it unable to fly off.

The second part of the mission was to deliver a member of the Expedition 19 crew, Koichi Wakata – Flight Engineer (Jpn), and return Expedition 18 crew member Sandra Magnus – Flight Engineer.

With all projects completed, the spacecraft Discovery returned to Earth on 28 March, landing at the Kennedy Space Center.

Crew: Lee Archambault – Commander (A); Dominic Antonelli – Pilot (A); Joseph Acaba – MS (A); Steven Swanson – MS (A); Richard Arnold – MS (A); John Phillips – MS (A).

The Soyuz TMA-14 flight to the ISS, which launched on 26 March 2009, transported the two other members of the Expedition 19 crew, Gennady Padalka – Commander (R) and Michael Barratt – Flight Engineer, as well as space tourist Charles Simonyi, on his second paying flight, to the Space Station. TMA-14 is the 101st manned flight of a Soyuz spacecraft; however, it was the 100th to launch and land *manned*, as Soyuz 34 was launched unmanned to replace Soyuz 32, which returned to Earth empty.

The Soyuz TMA-14 spacecraft will remain docked to the Space Station for the remainder of the Expedition 19 mission, to serve as an emergency escape vehicle.

Soyuz TMA-14 could be the final flight of a space tourist to the International Space Station. With the retirement of the Space Shuttle in 2010, and the extension of the crew of the station to six members, all Soyuz crew positions in the foreseeable future should be occupied by Expedition crews, at least until another spacecraft, most probably the Orion spacecraft, is able to service the International Space Station.

Simonyi returned to Earth aboard TMA-13 spacecraft with the Expedition 18 crew of Yuri Lonchakov – Commander (R) and Michael Finke – Flight Engineer, on 8 April 2009.

On 7 May an unmanned Russian M-O2M Progress cargo ship was launched to carry 2.5 tons of supplies, scientific instruments, food, water and newly designed spacesuits to the crew of the ISS. The new suits are completely computerised and warn the astronaut in the event of any problems when used during EVAs.

Launched from Kazakhstan, this was the second flight of the modified unmanned Progress spacecraft. The trip to the ISS, which normally takes two days, was scheduled to take five days in order to test all the systems after removing the faults discovered on the first flight in November 2008. The modernised Progress has a new main computer and features a fully digital telemetric system. The freighter is equipped with an on-board digital control system which replaces the analog-controlled Progress spacecraft.

After locking on to the Pirs port section of the ISS, the supply craft was unloaded by the resident crew of Gennady Padalka, Michael Barratt and Koichi Wakata. The spacecraft will stay connected to the ISS until discarded with unwanted experiments and rubbish, when it will be put into a decaying orbit to burn up on re-entry.

The next scheduled flight, to be the fifth and final servicing mission to the Hubble Space Telescope (HST), was STS-125 Atlantis (OV-104). The mission was flown with another shuttle ready to launch in case a rescue mission was needed.

Two Space Shuttles on the launch pad; the second is on stand-by in case there is an emergency during the mission of the other shuttle.

Due to an anomaly aboard the telescope that occurred on 27 September 2008, the launch of STS-125 was delayed until May 2009 to prepare a second data handling unit replacement for the telescope.

The launch on 10 May went smoothly and, once in orbit, preparations were made to close with the Hubble Space Telescope. On the third day in space the CandArm captured the Space Telescope and placed it in Atlantis' payload bay ready for servicing. The crew, Scott Altman – Commander, Gregory Johnson – Pilot, Michael Good – MS, Megan McArthur – MS, John Grunsfeld – MS, Michael Massimino – MS and Andrew Feustel – MS, were surprised at the remarkable condition of the Space Telescope after nineteen years in space.

Atlantis carried two new instruments to the HST: a replacement Fine Guidance Sensor and six new gyroscopes and batteries to allow the telescope to continue to function at least through 2013. The crew also installed a new thermal blanket layer to provide improved insulation, and a 'soft-capture mechanism' to aid in the safe de-orbiting of the telescope by an unmanned spacecraft at the end of its operational lifespan.

With all the servicing and repairs completed, the Hubble Telescope was placed back into its orbit by the CandArm and released. The mission is the last planned manned mission to the Space Telescope and was the most complex and dangerous to date. Three spacewalks of over 6 hours each were needed to complete the work; these were extremely arduous and dangerous tasks. Due to the difference

The Hubble Telescope in orbit after being serviced.

between the orbit of the ISS and that of the HST, Atlantis would be unable to reach the safe haven of the ISS in the event of its heatshield becoming damaged upon launch, or should any other major problem occur which prevented the spacecraft returning to Earth. Therefore, the mission required another shuttle – STS-400 (Endeavour) – to be ready on launch pad 39B for immediate flight on a Launch On Need (LON) rescue mission throughout the STS-125 mission. NASA contemplated whether it would be possible to use the same pad to launch both STS-125 and STS-400 (if needed), but it was decided to use two launch pads – 39A and 39B.

The landing was delayed for one day because of the unstable weather conditions at the Kennedy Space Center, and because of this Atlantis was diverted to the Edwards Air Force Base, the alternative landing site.

The crew of STS-400, Christopher Ferguson – Commander, Eric Boe – Pilot, Robert Kimbrough – MS and Stephen Bowen, are part of the STS-126 crew and were on standby throughout the mission.

The next spacecraft to visit the ISS was a Progress M-02M unmanned cargo craft. The spacecraft docked with the ISS on 13 May delivering 2.5 tons of supplies, including food, water and a number of scientific experiments. The trip to the Space Station took five days as a number of tests were carried out on the spacecraft. It was scheduled to undock from the ISS on 30 June but was used as a technical platform for a number of experiments.

The next ISS crew, Expedition 20, was launched aboard Soyuz TMA-15 on 27 May. This mission was to make history when, after docking with the ISS, it would be the first time the Space Station had a six-person crew. The Expedition 20 crew consisted of Roman Romanenko – Commander (R), Robert Thirsk (Can) and Frank de Winne (Belg), a truly international crew.

On 28 May 2009, NASA signed a $306 million agreement with the Russian Federal Space Agency for the training, preparation, launch and landing for American ISS crew members. This will cover four Soyuz missions in 2012; two in the spring and two in the autumn. This includes all medical examinations and services. The Soyuz spacecraft will be limited to carrying 110lb of cargo per person on launch, and 37lb of cargo and 66lb of trash per person on return.

Despite a lengthy delay because of a serious leak in the fuel tanks and adverse weather conditions, STS-127 Endeavour (OV-105) lifted off the pad at the fifth attempt on 16 July, the fortieth anniversary of Apollo XI. Once in orbit, the crew, Mark Polansky – Commander, Doug Hurley – Pilot, Dave Wolf – MS, Christopher Cassidy– MS, Julie Payette – MS (Can), Tom Marshburn – MS and Tim Kopra– MS, carried out an examination of the heatshield tiles using a camera on the end of the robotic arm, after reports that during the launch some

STS-127 crew patch.

pieces were seen to fly off the external tank. After intense examination nothing was found and the mission continued as planned.

Tim Kopra replaced Koichi Wakata and joined the Expedition 20 crew as Flight Engineer.

On Friday 17 July, Endeavour closed with the ISS and carried out a Rendezvous Pitch Manoeuvre (RPM), which meant that the spacecraft rolled on to its back allowing the crew of the ISS to take high-resolution photographs of the spacecraft's heatshield to ensure that no damage had been done. This manoeuvre took 9 minutes and was carried out about 600ft below the Space Station. With this successfully completed, Endeavour docked with the ISS and, after checking for leaks, the hatches were opened and the crew of the ISS welcomed their visitors.

Just before the STS-127 crew boarded the Space Station, two of the resident crew, Gennady Padalka and Frank de Winne, had to repair one of the three toilets in the station. Although a relatively simple job on Earth, because of the weightless environment, extra care had to be taken because the damaged pump had been flooded with a chemical liquid used in the plumbing process.

Over the past ten years, the development of the ISS has given it the living space of a jumbo jet. This enables the crew members to have a degree of privacy, which reduces tension when faced with a long-term mission in such a relatively close environment.

Three EVAs were carried out during the mission. The first two spacewalks were carried out without incident, but the third was cut short after carbon dioxide levels in Chris Cassidy's spacesuit started to rise unexpectedly. The two astronauts, Chris Cassidy and David Wolfe, were in the process of replacing the six batteries on the ISS' photovoltaic power system when the problem was discovered. The astronauts had replaced two of the batteries when they were recalled. It was decided to replace the remaining four batteries, which are designed to last six and a half years, on a later mission to the ISS. Once back inside the Endeavour, the problem was traced to the lithium hydroxide canister in the suit which was replaced.

The main mission for STS-127 was the installation of the Japanese Experiment Section to the Kibo laboratory. This required removing it from Endeavour's payload bay using the robotic arm, extending to a position where the ISS' robotic arm could grasp the payload and move it into place. Then two of Endeavour's astronauts, Chris Cassidy and Tom Marshburn, secured it to the Kibo laboratory.

With all tasks completed, the Endeavour's crew said their farewells to the crew of the ISS and began preparations to return to Earth. Koichi Wakata, having spent the last 133 days aboard the ISS, said his goodbyes and joined the crew of Endeavour. Two more small satellites were launched, a Global Positioning Satellite (GPS) and a satellite to measure the density of the Earth's rarefied atmosphere. After separating from the ISS, a final inspection of the spacecraft's heatshield was carried out and found to be okay.

The Orbiter Endeavour touched down at the Edwards Air Force Base on 31 July after completing fifteen days in space.

STS-128 Discovery (OV-103) lifted off from the Kennedy Space Center at 2359 hours on 28 August carrying the Space Station's next Expedition 21 crew member, Nicole Stott. The remaining members of the crew, Rick Sturckow (Commander), Kevin Ford (Pilot) with Mission Specialists Patrick Forrester, Jose Hernandez, Christer Fugslesang and John Olivas were tasked with bringing extra supplies and equipment to the station.

Amongst these was the Multi-Purpose Logistics Module which contained three racks for life support, a crew quarter to be installed in Kibo, a new treadmill to be placed temporarily in Node 2 and later in Node 3, and an Atmospheric Revitalisation Sysytem (ARS) to be placed temporarily in Kibo and later in Node 3. The module also contained racks dedicated to science, FIR (Fluids Integrated Rack) and the first Materials Science Research Rack (MSRR-1) to be placed in Desiny and a Minus Eighty Laboratory Freezer (MELFI-2) to be placed in Kibo. The FIR will enable detailed study of how liquids behave in microgravity, a crucial experiment for the many chemical reactions that constantly take place.

As Discovery approached the ISS Rick Sturckow rolled the Orbiter on to its back to allow the Space Station crew to photograph the spacecraft's heatshield. These inspections had taken place ever since Columbia had disintergrated on its return to Earth. Sturckow then carried out a straightforward docking with the ISS on 30 August bringing the number of astronauts aboard the Space Station to thirteen. It also brought together two astronauts from the European Space Agency: the Belgian Frank de Winne and Christer Fugslang, from Sweden.

One problem that was discovered was the failure of one of the two small manoeuvring jets mounted in the nose of Discovery. The crew closed a manifold that isolated and disabled both jets for the remainder of the mission. Mission Control confirmed that the loss of these two small jets would have no bearing on the mission and on the return to Earth.

The STS-128 crew took part in seat vibration tests that will help engineers on the ground understand how astronauts experience launch. They will then use the information to help design the crew seats that will be used in future NASA spacecraft.

STS-128 repeated the Boundary Layer Transition (BLT) Detailed Test Objective (DTO) experiment that was carried out by the same shuttle during STS-119. In this experiment, one of the thermal protection system tiles was raised to create a boundary layer transition in which the air flow became turbulent beyond a certain speed. During STS-119 the tile was raised 0.25in (6.4mm) above the others, tripping the flow at Mach 15 during re-entry. In the modification being done, the tile has been raised 0.35in (8.9mm) which will trip at Mach 18 producing more heat.

Discovery also undertook the testing of a catalytic coating which is intended to be used by the Orion spacecraft. Two TPS tiles located in the protuberance downstream from the BLT tile had been fully coated with the catalytic material in order to understand the entry heating performance. The tiles are instrumented to collect a wide variety of data.

With the transfer of the Leonardo Multi-purpose Logistics Module from Discovery's payload bay to the Harmony node completed, preparations were made to unload the contents into the ISS. These included a new crew quarters unit, a freezer for storing medical and experiment samples, food, supplies and a treadmill totaling 7.5 tons.

During one of the three spacewalks, Nicole Stott and Danny Olivas removed an empty ammonia tank from the ISS and stowed it in the payload bay of Discovery ready to be returned to Earth. A new ammonia tank was installed the following day by Danny Olivas and Christer Fuglesang.

With all the installations and checking completed the spacecraft prepared to undock from the ISS. After undocking, the spacecraft circled the Space Station to carry out a photographic survey of the ISS. Then, before making their final burn away from the Space Station, Discovery's crew used the robotic arm with a camera attached to check the spacecraft's heatshield.

The problem with space debris raised its head once again when a piece of debris of approximately 19 square metres in size was tracked in an elliptical orbit, became the cause of some concern because of its closeness to the ISS. Fortunately, at its closest it was 3km away and of no immediate threat. The piece of debris was later identified as a remnant of an Ariane 5 rocket. Landing at the Kennedy Space Center was delayed because of continuing bad weather so Discovery was diverted to land at Edwards Air Force Base in California.

Another cargo spacecraft arrived at the ISS on 18 September, only this time it was a Japanese HTC freighter carrying 4.5 tons of supplies. The unmanned spacecraft was captured by the recently installed Canadian robotic arm and was the first time it had been used by the crew of the ISS. The 10m-long cylindrically shaped space freighter cost $217 million, and, after being unloaded, was filled with waste and allowed to burn up as it entered the Earth's atmosphere. The spacecraft was developed at a cost of $800 million but was designed to be modified to carry a crew of three astronauts.

On 22 September the Progress M-67 cargo spacecraft that had arrived on 29 July undocked from the ISS. The Russian cargo spacecraft had carried 2.5 tons of supplies to the ISS and after being unloaded was re-filled with rubbish and unwanted experiments. After being released from the Space Station, the spacecraft was used as an unmanned orbital laboratory to conduct a number of geophysical experiments. After five days the cargo spacecraft was allowed to re-enter the Earth's atomosphere and to plunge into the Pacific Ocean. This was the last of the analogue-controlled Progress spacecraft as the new generation was to be digitally controlled.

The Expedition 21 crew blasted off from Baikonur on 30 September 2009 in their Soyuz TMA-16 spacecraft, with two replacement ISS crew members, astronaut Jeffrey Williams and cosmonaut Maxim Surayev. With them was fare-paying Canadian billionaire Guy Laliberte, the owner of the Cirque du Soleil, who is likely to be the last of the fare-paying passengers to go into space for some considerable time. This is because with the retirement of the shuttle, there will be very limited space available in the Russian Soyuz spacecraft which will be the only method of exchanging crews for the foreseeable future. When the spacecraft docked with the ISS it was the first time that there were three Soyuz spacecraft docked with the Space Station at the same time.

The Expedition 21 crew was joined by astronaut Nicole Stott who had arrived aboard STS-129. During Laliberte's stay on the Space Station he carried out a number of Earth-linked concerts to highlight the problems of water shortage in some parts of the world and the need to conserve it. Guy Laliberte returned to Earth on 11 October together with cosmonaut Gennady Padalka and astronaut Michael Barratt.

With the departure of Gennady Padalka, Frank De Winne took over as commander, the first European to hold the post. He was to be replaced in December by American astronaut Jeffrey Williams after spending six months aboard the ISS.

On 16 October the Russians launched the Progress M-03M unmanned cargo to ship to the ISS. The spacecraft contained 2.5 tons of of supplies which included food, water, fuel, some personal effects and scientific experiments. As per usual the empty supply craft was filled with trash, unwanted items and defunct scientific material. The cargo ship was then released and sent into a decaying orbit where it burnt up on re-entry.

A significant step in the development of a new spacecraft took place on 28 October when the Ares 1-X rocket lifted off the pad at Cape Canaveral, Florida. At 93m the Ares 1-X is the tallest rocket ever built and was designed to carry the newly designed Orion capsule that could carry six astronauts to the

ISS. However the Ares/Orion spacecraft will not enter service until 2015 at the earliest, so it will be left to the Russians and their Soyuz spacecraft to put the replacement crews on board the ISS. This will, of course, be heavily funded by the US government and NASA.

There are concerns within the United States that the budget for the Constellation programme, as it is known, is already spiralling out of control. The initial budget allocated for the project was $28 billion; this has now reached $44 billion. Critics say this is too much for the country to bear, especially during a time of recession, and a comment from the chairman of a commission set up to look at the cost of the Constellation programme said: 'The programme appears to be on an unsustainable trajectory and was attempting to reach goals not matched by its resources.'

The latest mission to the ISS at the time of writing lifted off from Cape Canaveral on 16 November 2009. STS-129 – Atlantis (OV-104) carried astronauts Colonel Charlie Hobaugh USMC (Commander) and Barry Wilmore (Pilot), accompanied by Mission Specialists Lt Col. Randy Bresnik USMC, Mike Foreman, Leland Melvin and Robert Satcher. It also carried 27,000lb of materials and supplies to the Space Station. Amongst the materials were replacement ammonia tanks and gyroscopes which are intended to help extend the life of the Space Station past 2010 when the ageing shuttle fleet will be retired.

After the routine examination of the tiles on the underside of the Orbiter Atlantis had been carried out, the spacecraft docked with the Space Station and the task of unloading all the supplies began. The following day Mission Specialists Mike Foreman and Robert Satcher began the first of the three planned spacewalks to install the equipment ready for the arrival of the Tranquility node when it arrives in 2010. This included a set of cables along the Destiny laboratory for a space-to-ground antenna due to arrive in 2010, and a bracket that will be used to help route an ammonia cable to the Tranquility node. A number of housekeeping duties were also carried including the lubricating of the latches on the ISS' mobile base system and replacing one of the communications antenna.

With their spacewalk completed and with 2 hours to spare, the two astronauts took the opportunity to deploy the Payload Attach System (PAS) on the outside of the Space Station's Starboard 3 truss. This will allow spare parts for the station's truss segments to be stored in a convenient position for future use.

The following day the whole station was awoken by alarms indicating depressurisation and the detection of smoke. It was quickly discovered that the alarms, coming from the Russian Poisk Mini-Research Module installed some weeks earlier, were false and the next 6-hour and 8-minute spacewalk was started by Mike Foreman and Randy Bresnik. During their mission, the two astronauts installed a cargo attachment system to the Starboard 3 truss together with a wireless video system that transmitted pictures back to the station and then relayed them to Earth.

The third and final spacewalk carried out by Randy Bresnik and Robert Satcher lasted 5 hours and 2 minutes, during which time they transferred a High Pressure Gas Tank from the Express Logistics Carrier 2 to a section of the Quest airlock. The kennel-shaped tank was moved with the help of the Space Station's robotic arm operated by Leland Melvin and Barry Wilmore. The tank will be used to replenish the air that is lost when spacewalkers exit and enter the airlock. Prior to this the two astronauts removed the airlock's debris shields to install the tank.

The crew of STS-129 joined the crew of the ISS to witness the formal handing over of command of the International Space Station to Jeffrey Williams. The outgoing commander, Frank de Winne, remained aboard the ISS until 1 December. During the mission, Randy Bresnik became a father for the first time when his wife Barbara gave birth to their daughter Abagail.

Atlantis undocked from the ISS on 25 November and circled to the Space Station for a final check. With this completed, Barry Wilmore fired the spacecraft's thrusters to move Atlantis well away from the Space Station before making preparations to return to Earth. Returning with the crew of STS-129 after their eleven-day mission was Nicole Stott, who had been the flight engineer aboard the Space Station since August 2009.

Atlantis returned to Earth on 27 November after a completing a perfect misson. Meanwhile, back on the Space Station, the ISS Expedition 21 crew of Frank de Winne, Roman Romanenko and Robert Thirsk, started their preparations to return to Earth. Their spacecraft, Soyuz TMA-15, undocked from the Space Station on 30 November, leaving just Jeffrey Williams and Max Suraev to man the ISS. The next Expedition 22 members of the crew did not board the ISS until 23 December 2009.

Concerns were once again voiced regarding space debris when the remnants of a Delta rocket, launched in February 1999, were spotted on radar moving close to the Space Station. The possibility of manoeuvring the ISS out of its path was avoided when it was realised that in fact the debris was moving away from the Space Station's orbit. Flight Director Dana Weigel decided against waking the crew of the Space Station after it was realised that the debris was estimated to be only 4in in diameter and over 9 miles away. The fact that the piece of debris was only small did not detract from the seriousness of the situation had the piece hit and punctured the outer skin of the ISS.

In the meantime, the Expedition 21 crew landed back on Earth and were met by Russian Search and Rescue Forces travelling in all-terrain vehicles. The helicopters that usually met incoming spacecraft were all grounded beause of low cloud and freezing temperatures. After being quickly extracted from the Soyuz spacecraft, the three crew were taken to Arkalyk where they were examined by doctors and found to be in excellent shape.

Talks are ongoing regarding extending the shuttle programme so as not to be reliant upon the Russian Soyuz spacecraft. Like everything connected with the space programme, it will inevitably come down to cost. NASA has already started planning for the shuttle's retirement by instructing manufacturers to halt production of the fuel tanks and Solid Rocket Boosters. This has inevitably resulted in hundreds of personnel being laid off, and such will be the impact on the US job market with the demise of the Space Shuttle.

The next shuttle mission is scheduled for February 2010 and, like the following four missions, will carry only supplies and scientific research equipment to the ISS. Nicole Stott, who returned aboard STS-129, will be the last American to make the round trip to the ISS and back aboard the shuttle.

CHAPTER TEN

THE END OF THE SPACE SHUTTLE

The final Space Shuttle crew has been selected. The launch of STS-133 is scheduled to take place in September 2010. The crew, Steven Lindsey – Commander; Eric Boe – Pilot; Alvin Drew – MS; Michael Barrat – MS; Timothy Kopra – MS; and Nicole Stott – MS, are all experienced astronauts having carried out previous spaceflights.

With the phasing out of the Space Shuttle there are even thoughts about reviving the Russian Space Shuttle Buran programme. One of the main differences between the Russian and American Orbiters was that the Buran could be landed automatically, unlike its American counterpart, which was landed manually. As with the development of the Space Shuttle, the cost of re-developing the Buran into a viable concept would be astronomical.

One of the major problems facing the space scientists today is the ever increasing amount of space debris that is being dumped in the Earth's orbit with almost every launch of manned spacecraft or satellites. Some of the early rocket stages that remain in orbit are slowly coming apart and adding to almost 600,000 pieces in the junkyard that is now cluttering the space around the Earth. In May 2009 the third stage of a Russian Tsyklon rocket exploded into more than fifty pieces, most of which were large enough to be tracked by ground-based sensors. In June a Russian Proton rocket motor, which was launched in the 1980s, broke up and went into an elliptical orbit with a perigee of 655km and an apogee of 18,410km. Recently the remnants of a Chinese rocket narrowly missed the Hubble Telescope, and the crew of STS-125, whilst servicing the telescope, were very aware of the pieces of junk that were spinning around the Earth in their orbit at the time. On 10 February 2009 a defunct Russian military satellite collided with the Iridium 33 communication satellite, destroying both objects and sending 700 pieces of debris to join the ever-expanding 'junkyard'.

If the problem is not addressed soon, spaceflights to the ISS and beyond are going to have to be put on hold until the debris field that is being created is cleared, as it will become increasingly dangerous as more and more satellites and their rockets are launched and become defunct.

With the last of the shuttle flights scheduled to be completed in 2010, the way is now open to a new form of space transportation – Project Constellation. With further development of the ISS now first and foremost in the minds of the Western powers, the Orbiter spacecraft appeared to be the ideal way of transporting the materials and personnel into orbit in order to build them. As time went on and the Space Station programme developed, the use of the Orbiter became more and more frequent with the intention of phasing out the rocket programme. Then came two disasters concerning the Space Shuttle and all missions were grounded. This left only the Russians with their manned and unmanned Soyuz spacecraft to keep the ISS supplied with the necessary replacement crews and supplies. The shuttle, however, is the only spacecraft capable of carrying heavy, bulky equipment into space and to the Space Station, so keeping the ISS operational until 2020 could be a problem.

But what of the future? The X-33 and X-34 spaceplanes, the forerunners of the Venture Star, had been on the drawing boards and prepared for evaluation. These completely reusable aircraft/spaceplanes were designed to take off from a conventional runway then, just before reaching the atmosphere, shut down their jet engines, ignite their rocket engines and blast their way into space. Their return was to be the same as that of the present Space Shuttle Orbiter. But because of numerous problems with the spaceplanes, the projects were finally cancelled. This meant that an alternative programme had to be found – Project Constellation.

The completion of the ISS is scheduled to be finished in 2010 and to coincide with the phasing out of the Space Shuttle. A new type of spacecraft, the Orion Crew Exploration Vehicle (CEV) that will ferry supplies and crew members to and from the ISS, will be built. But the main object is to return man to the Moon with the intention of establishing a base there. Future flights are already being planned to go to Mars and it is intended that the Moon base will be the launch base for such missions.

To prepare for this a series of unmanned robotic missions will be sent to the Moon with the intention of sending a manned exploration mission in 2015. The idea behind the setting up of a base on the Moon is to assemble spacecraft that could be launched using far less fuel because of the low gravity, and therefore reduce the costs considerably.

Two types of booster rockets have been designed, Ares I and Ares V. Ares I will be used to place crews into orbit, whilst Ares V will have a heavier lift capability, and be used to launch a variety of hardware that will be used for other missions.

Co-operation between the major powers and the sharing of information and technology is slowly unlocking the doors to the universe. Who knows what lies beyond?

APPENDIX

American Rockets

Redstone (Mercury-Redstone Missions)

Height	68ft 11in (21.0m)
Diameter	5ft 11in (1.8m)
Engines (Single Stage)	Rocketdyne North American Aviation 75-110A-7
Launch Weight	62,710lb (28,440kg)
Propellant	Lox/Alcohol

Atlas-D (Mercury-Atlas Missions)

Height	71ft 3in (21.7m)
Height Overall (Shroud)	87ft 0in (26.5m)
Diameter	10ft 3in (3.0 m)
Engines (Single Stage)	One Rocketdyne MA-3 1½ stage system consisting of two boosters (165,000lb thrust), one sustaining engine (57,000lb thrust) and two vernier engines (1,000lb thrust). The total thrust output as 389,000lb. The two booster engines were jettisoned in flight
Launch Weight	269,000lb (122,000kg)
Propellant	Lox/Kerosene

Titan II (Gemini Missions)

Height Overall (with capsule)	118ft (36.0m)
Height (First Stage)	73ft (22.3m)
Height (Second Stage)	25ft 11in (7.9m)
Diameter (Overall)	10ft (3.1m)
Diameter (First Stage)	10ft (3.1m)
Diameter (Second Stage)	10ft (3.1m)
Engines (First Stage)	Thrust 430,000lb
Engines (Second Stage)	Thrust 100,000lb
Fuel	N204/Aerozine
Weight (First Stage)	259,870lb (117,855kg)
Weight (Second Stage)	63,810lb (28,939kg)

Saturn 1B

Height Overall (with Apollo Capsule)	224ft 0in (68.2m)
Height	173ft 0in (56.2m)
Height (First Stage)	80ft 4in (32.6m)
Height (Second Stage)	58ft 5in (23.7m)
Diameter (First Stage)	21ft 5in (6.5m)
Diameter (Second Stage)	21ft 8in (6.6m)
Launch Weight (First Stage)	882,000lb (400,075kg)
Launch Weight (Second Stage)	230,000lb (104,328kg)
Launch Weight	1,300,000lb (589,680kg)
Engines (First Stage)	Eight Rocketdyne H-1 and four outboard gimballed engines giving 205,000lb thrust each
Fuel (First Stage)	Liquid Oxygen/Kerosene
Engines (Second Stage)	One Rocketdyne J-2 restartable engine giving 215,000lb thrust
Fuel (Second Stage)	Liquid Oxygen/ Liquid Hydrogen

Saturn V (Apollo Missions)

Height Overall (with Apollo Capsule)	363ft 0in (110.6m)
Height	281ft 0in (85.6m)
Height (First Stage)	138ft 0in (42.1m)
Height (Second Stage)	81ft 6in (24.8m)
Height (Third Stage)	59ft 4in (18.1m)
Diameter (First Stage)	33ft 0in (10.1m)
Diameter (Second Stage)	33ft 0in (10.1m)
Launch Weight (First Stage)	4,952,755lb
Launch Weight (Second Stage)	1,087,580lb
Launch Weight (Third Stage)	24,900lb

Engines (First Stage)	5 Rocketdyne F-1 and four outboard gimballed engines giving 1,533,022lb each
Fuel (First Stage)	Liquid Oxygen/Kerosene
Engines (Second Stage)	Five Rocketdyne J-2 engines with gimballed nozzles on four outer engines giving 230,740lb thrust
Fuel (Second Stage)	Liquid Oxygen/Liquid Hydrogen
Engines (Third Stage)	One Rocketdyne J-2 restartable engine giving 215,000lb thrust
Fuel	Liquid Oxygen

American Spacecraft

Mercury (One-man Crew)

Height	9ft 6in (2.9m)
Height (with Escape Tower)	16ft (4.9m)
Diameter	6ft 2in (1.88m)
Orbital Mass	3,027lb (1,376kg)

Gemini (Two-man Crew)

Height	18ft 5in (5.6m)
Height (Re-entry Module)	11ft (3.25m)
Diameter Base	10ft (3.0m)
Diameter (Re-entry Module – Top)	3ft 3in (1.0m)
Diameter (Re-entry Module – Base)	7ft 6in (2.3m)
Orbital Mass	8, 297lb (3,763kg)

Apollo VII, VIII, IX, X, XI, XII, XIII, XIV and Skylab (Three-man Crew)

Height (Command Module)	10ft 7in (3.2m)
Height (Service Module)	24ft 7in (7.5m)
Height Overall	34ft 7in (10.3m)
Diameter	12ft 10in (3.9m)
Weight (Launch)	63,493lb (28,800kg)
Weight (Service Module)	51,243lb (23,243kg)
Weight (Empty SM)	13,450lb (6,100kg)

Propellant

SPS Fuel	15,690lb (7,107kg)
SPS Oxidizer	25,106lb (11,373kg)
RCS	1,342lb (607kg)

Apollo – J Model XV, XVI and XVII (Three-man Crew)

Height (Command Module)	10ft 7in (3.2m)
Height (Service Module)	24ft 2in (7.4m)
Height Overall	34ft 9in (10.3m)
Diameter	12ft 10in (3.9m)
Weight (Launch)	66,844lb (30,320kg)
Weight (Service Module)	54,055lb (24,514kg)
Weight (Empty SM)	13,450lb (6,100kg)

Propellant

SPS Fuel	15,690lb (7,107kg)
SPS Oxidizer	25,106lb (11,373kg)
RCS	1,342lb (607kg)

Lunar Excursion Module

Lunar Module

Height (legs extended)	22ft 11in (6.9m)
Diameter (across extended legs)	31ft (9.4m)

Ascent Stage

Height	12ft 4in (3.7m)
Diameter	14ft 1in (4.2m)

Descent Stage

Height	10ft 7in (3.2m)
Diameter	14ft 1in (4.5m)
Weight (Overall with Crew)	32,400lb (14,677kg)
Ascent Stage	5,200lb (2,355kg)
Descent Stage	18,000lb (8,154kg)

Lunar Roving Vehicle (LRV)

Weight	210kg (462lb) on Earth, 34kg (76lb) on Moon
Length	10ft 2in (3.1m)
Width	6ft 0in (1.8m)
Height	3ft 8in (1.14m)
Wheelbase	7ft 6in (2.3m)
Turning Radius	10ft 0in (3m)
Drive Power	One ¼-hp motor of 10,000rpm attached to each wheel
Power Source	Two 36 volt, silver-zinc batteries
Payload	1,080lb (490kg)
Crew	Two

A total of four LRVs were built at a total cost of $38 million, three were taken and used on the Moon, the fourth was used in tests and is currently on display at the NASM (National Air and Space Museum).

Skylab

Length (Overall)	117ft (35.6m)
Diameter (Overall)	90ft (27.4m) (including solar array)
Weight (including the CSM)	199,750lb (90.60kg)

Orbital Workshop

Length	48ft (14.4m)
Diameter	22ft (6.6m)
Volume	9,550 cubic feet (270 cubic metres)

Multiple Docking Adaptor

Length	17ft (5.1m)
Diameter	10ft (3.0 m)
Volume (Habitable)	1,080 cubic feet (32.40 cubic metres)

Space Shuttle Orbiters

Length	122ft 0in (37.19m)
Wing Span	78ft 0in (23.77m)
Height	57ft 0in (17.37m)

Challenger	OV-099
Enterprise	OV-101
Columbia	OV-102
Discovery	OV-103
Atlantis	OV-104
Endeavour	OV-105

External Tank

Length	154ft 0in (46.93m)
Diameter	27.5ft 0in (8.38m)
Engines	Three main engines plus two OMS (Orbital Manoeuvring System) engines. The main engines producing 375,000lb thrust each
Fuel (Liquid Hydrogen)	378,378 gal (1,432,161 litres)
Fuel (Liquid Oxygen)	139,623 gal (528,473 litres)

Solid-fuel Rocket Boosters

Length	149ft 0in (45.42m)
Diameter	12ft 0in (3.66m)
Fuel	The propellant mixture in each SRB motor consists of ammonium perchlolate (oxidiser 69.6% by weight), aluminium fuel (16%), iron oxide(a catalyst 0.4%), a polymer (such as PBAN or HTPB, serving as a binder that holds the mixture together and acting as secondary fuel, 12.04%), and an epoxy curing agent (1.96%). This propellant is commonly referred to as Ammonium Percholate Propellant, or simply APCP.
Propellant Weight	1,106,000lb (501,681kg) each
Crew	Two on first four flights, up to seven on later missions

List of shuttle flights that required a rescue flight to be at readiness:

Mission	*Rescue Shuttle*
STS-114 (Discovery)	STS-300 (Atlantis)
STS-121 (Discovery)	STS-300 (Atlantis)
STS-115 (Atlantis)	STS-301 (Discovery)
STS-116 (Discovery)	STS-317 (Atlantis)
STS-117 (Atlantis)	STS-318 (Endeavour)
STS-118 (Endeavour)	STS-332 (Discovery)
STS-120 (Discovery)	STS-320 (Atlantis)
STS-122 (Atlantis)	STS-323 (Discovery)
STS-123 (Endeavour)	STS-324 (Discovery)
STS-124 (Discovery)	STS-326 (Endeavour)
STS-125 (Atlantis)	STS-400 (Endeavour)
STS-134 (Discovery)	STS-335 (Atlantis)

Russian Rockets

Vostok A-1

Height	124ft 7in (38.0m)
Boosters	62ft 4in (19.0m)
First Stage	90ft 2in (27.5m)
Second Stage	8ft 6in (2.6m)
Shroud	8ft 9in (2.8m)
Diameter (Base)	33ft 9in (10.3m)
Diameter (Booster)	10ft 0in (3.0 m)
Diameter (First Stage)	9ft 7in (2.95m)
Diameter (Second Stage)	8ft 6in (2.6m)
Engines (First Stage)	One RD 108 consisting of four thrust chambers and four verniers giving 224,910lb thrust on lift-off
Engines (Second Stage)	A single chamber giving 198,450lb thrust for second 'burn'
Engines (Booster)	Four boosters each with a single RD 107 engine consisting of four thrust chambers and two verniers giving 224,910lb thrust per booster

Soyuz A-2

Height	167ft 4in (51.0m)
Boosters	62ft 4in (19.0m)
First Stage	90ft 2in (27.5m)
Second Stage	30ft 9in (30.8m)
Shroud	8ft 9in (2.8m)
Diameter (Base)	33ft 9in (10.3m)
Diameter (Boosters)	10ft 0in (3.0m)
Diameter (First Stage)	9ft 7in (2.95m)
Diameter (Second Stage)	8ft 6in (2.6m)
Diameter (Shroud)	8ft 9in (2.8m)
Engines (First Stage)	One RD 108 consisting of four thrust chambers and four verniers giving 211,640lb thrust
Engines (Second Stage)	One four-chamber engine giving 66,138lb thrust
Engine (Booster)	Four boosters with a single RD 107 engine consisting of four thrust chambers and two verniers giving 224,868lb thrust each booster

Russian Spacecraft

Vostok (One-man crew)

Height (with Booster Stage)	24ft 1in (7.35m)
Height (without Booster Stage)	15ft (4.6m)
Diameter	8ft 5in (2.58m)
Diameter – Capsule	10ft 6in (2.3m)
Launch Weight – Capsule	5,280lb (2,400kg)

Voskhod

Said to be almost identical to Vostok but with additional seating.

Soyuz (Two/Three-man crew)

Height Overall	23ft 4in (7.1m)
Height (Service Module)	9ft (2.7m)
Height (Descent Module)	7ft 2in (2.2m)
Height (Orbital Module)	8ft 7in (2.65m)
Diameter (Overall)	8ft 5in (2.58m)
Diameter (Service Module)	9ft 0in (2.7m)
Diameter (Descent Module)	7ft 3in (2.2m)
Diameter (Orbital Module)	7ft 3in (2.2m)
Launch Weight (Overall)	14,750lb (6,690kg)
Launch Weight (Service Module)	5,850lb (2,654kg)
Launch Weight (Descent Module)	6,200lb (2,812kg)
Launch Weight (Orbital Module)	2,700lb (1,224kg)

Progress (Unmanned)

Height	25ft 11in (7.9m)
Diameter	8ft 10in (2.7m)
Launch Weight (Overall)	15,479lb (7,020kg)
Payload Weight	5,071lb (2,300kg)

Space Stations

Salyut

Orginally called Zarya but renamed Salyut, the Space Station was launched on 19 April 1971. The cylindrical Space Station was 12m long by 4.1m wide, had two solar panels mounted laterally in the centre and two docking ports at either end. There were six Salyut stations launched during the following eleven years, each one being allowed to go into a decaying orbit. Salyut 3 lasted six months, Salyut 4 one year, Salyut 5 two years, Salyut 6 five years and Salyut 7 nine years.

Mir

This was the replacement for the Salyut Space Station and consisted of six connected modules. The main section, which was 43ft (13.1m) long and 14ft (4.2m) in diameter, was contrived from the Salyut 6 and 7 Space Station and contained the command centre, eating area and sleeping area for two cosmonauts. Attached to this was the Kvant module where the majority of waste was stashed. The Kvant module was 42.6ft (13.0m) in length and 14ft (4.2m) in diameter. Another of the modules was Kvant 2, which was 44.9ft (13.7m) in length and had a diameter of 14ft (4.3m). This was where the Space Station's main toilet was housed, amongst other things. Also attached was the Kristall module. This 39ft (11.8m) long module had a diameter of 14ft (4.2m) which was in keeping with the other modules. In 1995 an additional module was added – Spektr. This was used primarily by the visiting American astronauts during their stay aboard the Space Station and contained a variety of scientific equipment. This was followed later by Piroda, a 42ft 6in (13.0m) long by 14ft 2in (4.4m) diameter module.

International Space Station (ISS)

1st segment	Zaraya (Russia)
2nd segment	Unity (USA)
3rd segment	Zvezda (Russia)
4th segment	Destiny (USA)
5th segment	Quest (USA)
6th segment	Pirs (Russia)
7th segment	Harmony (Europe/USA)
8th segment	Colmbus (Europe)
9th segment	Kibo (Japan – First section)
10th segment	Kibo (Japan – Second section)
11th segment	Poisk (Russia)

Sections scheduled to be attached in 2010:

12th segment	Tranquility (USA)
13th segment	Cupola (Europe/USA)
14th segment	Rassvet (Russia)
15th segment	Leonardo (Europe/USA)
16th segment	Nauka (Russia)

GLOSSARY

ALSEP	Apollo Lunar Surface Experiments Package
APU	Auxiliary Power Unit
ATLAS	Atmospheric Laboratory for Applications
ASTP	Apollo Soyuz Test Project
CRISTA-SPAS	Cryogenic Infrared Spectrometers and Telescopes for the Atmosphere-Shuttle Pallet Satellite
CM	Command Module
CSM	Command and Service Module
EDT	Eastern Daylight Time
EST	Eastern Standard Time
EURECA	European Space Agency's European Retrievable Carrier
FRR	Flight Readiness Review
GMT	Greenwich Mean Time
HHMU	Hand Held Manoeuvring Unit
IML	International Microgravity Laboratory
INTELSAT	International Satellite
ISS	International Space Station
JPL	Jet Propulsion Laboratory
KGB	*Komitet Gosudarstvennoi Bezopasnosti*
LAGEOS	Laser Geodynamic Satellite
LEM	Lunar Excursion Module
LLRV	Lunar Landing Research Vehicle
LRV	Lunar Roving Vehicle
MAPS	Measurement of Air Pollution
MPLM	Multi-Purpose Logistics Module
MMU	Manned Manoeuvring Unit
MSC	Manned Space Center
NAA	North American Aviation
NACA	National Advisory Committee for Aeronautics

NASA	National Air and Space Administration
NSDA	National Space Development Agency of Japan
OAST	Office of Aeronautics and Space Technology
OPF	Orbiter Processing Facility
OSS	Office of Space Science
OV	Orbiter Vehicle
PST	Pacific Standard Time
PDT	Pacific Daylight Time
RCS	Reaction Control System
RPM	Rendezvous Pitch Manoeuvre
SARE	Shuttle Amateur Radio Experiment
SECO	Sustainer Engine Cut-Off
SM	Service Module
SPARTAN	Shuttle Point Autonomous Research Tool for Astronomy
SRB	Solid Rocket Booster
SRL	Space Radar Laboratory
STS	Space Transportation System
TDRS-G	Tracking and Data Relay Satellite-G
UARS	Upper Atmosphere Research Satellite